RANGERS, SCOUTS, AND RAIDERS

Origin, Organization, and Operations of
Selected Special Operations Forces

MICHAEL F. DILLEY

Philadelphia & Oxford

Published in the United States of America and Great Britain in 2023 by
CASEMATE PUBLISHERS
1950 Lawrence Road, Havertown, PA 19083, USA
and
The Old Music Hall, 106–108 Cowley Road, Oxford OX4 1JE, UK

Copyright 2023 © Michael F. Dilley

Hardcover Edition: ISBN 978-1-63624-283-5
Digital Edition: ISBN 978-1-63624-284-2

A CIP record for this book is available from the British Library

All rights reserved. No part of this book may be reproduced or transmitted in any form or by any means, electronic or mechanical including photocopying, recording or by any information storage and retrieval system, without permission from the publisher in writing.

Printed and bound in the United Kingdom by CPI Group (UK) Ltd, Croydon, CR0 4YY

Typeset in India by Lapiz Digital Services, Chennai.

For a complete list of Casemate titles, please contact:

CASEMATE PUBLISHERS (US)
Telephone (610) 853-9131
Fax (610) 853-9146
Email: casemate@casematepublishers.com
www.casematepublishers.com

CASEMATE PUBLISHERS (UK)
Telephone (01865) 241249
Email: casemate-uk@casematepublishers.co.uk
www.casematepublishers.co.uk

Computer graphic chart of the First Special Service Force designed by Dave Shroyer.

Organization charts and the insignia pictures are by the author; insignia are from the author's collection, except for the Glider Badge and the Navy and Marine Corps Parachutist Insignia, which are from a different collection.

Front cover: V-42 combat knife. (Wikimedia Commons)

Dedicated to the Rangers, Scouts, Raiders, Paratroopers,
Submariners, and Special Operators of the United States
Past, present, and future
With great respect

Contents

Acknowledgements		ix
Foreword		xi
Introduction		xv
1	The Return of Rogers' Rangers	1
2	The Andrews Raid	11
3	American Airborne Units in World War II	21
4	The Force at la Difensa	43
5	Sabotaging Hitler's Heavy Water	55
6	The Alamo Scouts—LRRPs of World War II	61
7	Special Allied Airborne Reconnaissance Force	71
8	Gypsy Task Force at Aparri	81
9	Recondo Training—Its Origins and Aftermath	91
10	The Son Tay Raid	99
Appendices		125
Bibliography		171
Index		175

Acknowledgements

Writers are rarely able to work and get their work published on their own. There are always others who have helped along the way. My writing is no different.

I owe my biggest writing thanks to Gary and Mike Linderer, the founders and publishers of *Behind The Lines* magazine, who gave my first writings a big boost by publishing my articles and book reviews in that excellent, but now sadly gone, magazine. During the years the magazine was published, Gary raised my status from Contributing Writer to Staff Writer and Senior Editor.

Over the years there have been others who have helped me in the writing process. There have been other writers as well as editors and publishers. Ray Merriam at Merriam Press, who published my first book, *Galahad*. Sam Southworth and Steve Smith at Sarpedon Publishers are high on that list. They approached me to take part in an anthology (*Great Raids in History*) that helped boost both my confidence as well as my name. Steve Smith was a big help in getting one of my other books (*Behind The Lines*) published at Casemate Publishers and has been a continuous help since. Dan Cragg and Jack Schaffer have been supporters of my writing for a very long time and I want to thank them as well.

I also must thank Lance Q. Zedric, my co-author on *Elite Warriors*. We both helped each other to get things published. Lance has been a good friend over the years and helped with several factual questions I have had about my research. In this same vein, I want to thank Russ Blaise of the Alamo Scouts Historical Foundation for his invaluable assistance.

Brian Williams has been helpful in assisting me get some items published. I appreciate his support.

A big thank you to the folks at Casemate for all their work in getting this work to completion, especially to Noelle Marasheski, Ruth Sheppard, Isobel Fulton, Felicity Goldsack, Eduard Cojocaru, Lizzy Hammond and Declan Ingram for all of their specific work.

In researching all that I have written I owe a great debt of gratitude to the many librarians, archivists, and research assistants who helped me over the years to find books and other reference material that I needed to complete my work and make it better. Among these folks are: Beverly McMaster, Research Division, Donovan Technical Library, Fort Benning, Georgia; Karla Norman, Research Department, Fort Leavenworth, Kansas; Roxanne Merritt, Research Specialist for the John F. Kennedy Special Warfare Museum, Fort Bragg, North Carolina; Delores E. Oplinger, U.S. Army Signal Corps Museum, Fort Gordon, Georgia; Rich Boylen, Archive Research Specialist, U.S. National Archives, Suitland, Maryland; Terry Van Meter, Chief of the Museum Division at Fort Riley, Kansas; and Pat Tugwell, Pentagon Library in Washington, D.C. If there are others that I may have not shown, I apologize.

Thanks to Dr. John Partin, command Historian, U.S. Special Operations Command, MacDill Air Force Base, Florida, for help in tracking down facts; to Nancy Guier and Roxanne Merritt, John F. Kennedy Special Warfare Museum, Fort Bragg, North Carolina for the photographs; and to the website for the U.S. Army's Quartermaster Museum, Fort Lee, Virginia, for the official descriptions of the parachutist's and glider wings.

Finally, but never last, I want to thank my wife and my best friend, Sue, for her help over the years. She has taken the time to edit all my work and to give me suggestions on how to make them all better. Whatever I have written would not be as good without her assistance.

Foreword

In 1964, when Mike Dilley enlisted in the U.S. Army, he hoped to gain enough training and experience to qualify as a police detective. He also wanted to write, and the fruit of that ambition is in *Galahad, Behind the Lines,* and many articles on special operations. Twenty years later as a major and an experienced intelligence professional, Mr. Dilley could have started a second career as Sherlock Holmes. Detectives and intelligence agents have a lot in common. And for that matter, so do historians: asking questions, sifting facts, cautiously building cases and weighing evidence—and the best ones never make snap judgments or rush to conclusions.

So *Rangers, Scouts, and Raiders* has been in the works for many years, the cumulative product of a soldier-historian's tribute to that gallant breed of men who have distinguished themselves, often far behind the lines. The scope is inclusive. Units that performed special missions, whether in the British Army prior to the American Revolution or Union soldiers in our Civil War, will be found here. Mr. Dilley follows the definition of special purpose, special mission organizations he developed with his co-author, Lance Q. Zedric in their book *Elite Warriors,* and it is worth reviewing here (see Introduction).

This is in many ways the very "bones" of special operations as they have been organized and used over the years. The scope is as broad as good military planning itself. Personal reminiscences are fascinating, useful primary sources for historians but what Mr. Dilley has given us here is a *record* gleaned from many sources.

As any student of History 101 knows, notes and bibliography are crucial. The bibliography alone contains the meat for generations of military historians. Mr. Dilley has read all those books and articles. But by far the most spectacular aspect of this book are the full-color reproductions of selected shoulder-sleeve insignia of the special operations units discussed in this book. They are from Mr. Dilley's personal collection. "Patch" collectors will gasp with delight. So did I, because these insignia embody the very spirit of the men who wore them, the "unauthorized" ones in particular because they show us how the men who designed them thought of themselves, not some military heraldry department, and how they wished to be remembered. They put the "blood" into the meat on the bones of Mr. Dilley's organizational history. An old soldier like Mr. Dilley knows how important these insignia are to the morale of the men who qualified to wear them. Those men do not speak directly to us here but these patches do, more eloquently than mere words ever could.

The value of this work can only be appreciated if you know something about how difficult it was. I think Mike Dilley is in complete agreement with that great American historian, William H. Prescott, who concluded "It is seldom safe to use anything stronger than *probably* in history."(*History of the Conquest of Mexico*, Vol. II, p. 144, n. 13 (NY, Harper & Bros., 1843)).

I am often reminded that the historian is like the little boy who was asked if he knew the words to the *Battle Hymn of the Republic*, and he replied, "Sure! 'He is trampling out the village where the great giraffe is stored!'" That's what misperception can do to memory. And then there are those in authority who have no sense of history. I knew a military historian in Vietnam who compiled a series of interviews with officers who'd served as advisors to the Vietnamese Army. Their memories and opinions were fresh and vivid and he included them, warts and all, in his paper. When he submitted it for approval he was told it was "worthless, too opinionated," and it should be destroyed. To an historian today that paper would be a golden fleece.

Published books and oral histories have their pitfalls too.

When old soldiers reminisce, as honest as they may try to be, how accurate are those memories, contaminated with years of experience in the interval? That's where the "giraffes" can sneak in. How do you find the details on a unit in war? Are they on microfilm? Digitized? Do you want to sit for hours scanning stuff whirling by on a computer? Mike Dilley has and he can tell you every now and then, just as you're ready for seasickness pills, a little gem pops out and by golly, you've got what you've been looking for! There is no greater happiness for a researcher. Then there are secondary sources, books written by great historians, people you can trust, until some reviewer (there's always *somebody* who knows *everything* about *something*) points out that your description of Caesar's uniform was wrong—the other guy's mistake, but you took it verbatim because you trusted him and you have the sinking feeling that one day some poor student will be misled by somebody else's mistake published under *your* name. Well, Mr. Dilley was never rushed in the preparation of this material, you can bet on that.

So, collections aren't complete, what is has to be sifted carefully, or what you really need just doesn't exist anymore. Mike knows this as well as anyone, so he picked his sources judiciously, such as that Virginia gentleman, the late Mayo Stuntz of the Alamo Scouts, and the late, great Nick Rowe, a Special Forces icon with whom he worked developing SERE (Survival, Evasion, Resistance and Escape) training for the 82nd Airborne Division. In addition to his own personal library, he consulted public and private collections such as the National Archives, where you sit hour after hour, day after day, going through old records boxes, winnowing the wheat from the chaff. Mr. Dilley is a veteran with the dedication and determination expected of anyone who would voluntarily jump out of perfectly good airplanes.

You will find no "giraffes" in Mike Dilley's work.

Daniel J. Cragg, Sergeant Major (Retired), U.S. Army
Co-author with Michael Lee Lanning of *Inside the VC and the NVA;*
Top Sergeant with Sergeant Major of the Army Bill Bainbridge;
A Dictionary of Soldier Talk, with John R. Elting and Earnest Deal.

Introduction

Throughout history there has always been a need, in military forces, for special units. In the past, these units have usually been *ad hoc* formations that were disbanded after their war or mission was over. It has only been in recent years, pretty much since the early 1950s, that such units have remained active. That doesn't mean that all units that have been raised since then have survived to the present—only some of them. Almost all of the units discussed in this book are of the latter kind; they were organized and were used for a period of time or for a specific mission and then either deactivated or replace by other units.

In a book that I co-wrote with Lance Q. Zedric (*Elite Warriors*), we came up with a definition of what we call special purpose, special mission units. Our definition is a bit broader than that used by most military forces but is an appropriate one. Since all of the units discussed in this book fall into the category of special purpose, special mission units, it is worthwhile repeating the definition here. A special purpose, special mission unit is an organization that has one or more of these qualifiers:

- Conducts missions atypical of units in its branch of service;
- Is formed to conduct a particular mission;
- Receives special training for a mission;
- Uses specialized or prototype equipment or standard equipment in a non-standard role;
- Performs scouting, ranging, raiding, or reconnaissance missions;
- Conducts or trains indigenous people in guerrilla type or unconventional warfare operations;
- Results from separate recruiting efforts, either in-service or off-the-street.

The units discussed in this book cover a period from the French and Indian War in the 18th century to the Vietnam War in the 20th century. I have selected these operations for their diversity in both their organization and their purpose, but also to show the continuous need for special units throughout recent history. They perform a variety of different functions. Another reason that I selected these operations is because the headquarters which controlled these units used them properly. Misuse of special units has been one of the problems of their existence. Another reason for including these operations is because they all contributed to the overall mission of their theater or larger organization, even in the one case when the mission failed. This also has not always been the case with special units in the past.

An appendix is included at the end of the discussions of the operations. This appendix is presented as a guide to the origins and development of American special purpose, special missions organizations beginning in 1670 with the formation of the first Ranger organization in the American colonies, although individual Rangers can be traced to 1622.

Finally, I commend these operations to you because I believe they are among the best conducted by special units.

CHAPTER I

The Return of Rogers' Rangers

The military exploits of Major Robert Rogers during the French and Indian War are well known. It was during that war that Rogers raised, trained, and led the unit that bears his name, Rogers' Rangers. This was, however, not the last Ranger unit with which Robert Rogers was affiliated.

Prior to the war, Rogers had narrowly missed being branded or hung as a result of a charge of counterfeiting. His exploits during the war left him with money problems but of a different nature. The new problems involved Rogers' accounts in the army—repaying some remaining obligations to his former Rangers as well as to certain men in Albany, New York, who had loaned money for the Rangers' subsistence, and loans some of the Rangers had taken against their future pay. Rogers spent almost a month preparing his statement and presented it to the Crown's representative. By his account the Crown owed Rogers about 6,000 pounds. Rogers was reported to have been "thunderstruck" when most of the statement he submitted was denied. Without detailing Rogers' claim for repayment by the Crown (or even the convoluted method of financing the British Army in the mid-1700s), suffice it to say Rogers went to his grave still being pursued by his creditors. His attempts to pay these creditors drove almost everything he did in life after the French and Indian War.

Rogers continued to serve as an active officer in South Carolina and in Detroit, the latter during what was called "Pontiac's Conspiracy," an almost last-gasp struggle by "northeastern woodland Indians against

Major Robert Rogers, commander of Rogers' Rangers. (Library of Congress)

white supremacy." It was in Detroit in 1763 that Rogers was one of 14 officers who preferred charges against Captain Joseph Hopkins, commander of a unit known variously as Hopkins' Rangers, the Queen's Royal American Rangers, and the Queen's Rangers. Hopkins was accused, among other things, of overcharging the men under his command for necessities furnished by him. In the end, though, Hopkins was not court-martialed.

During the next 12 years Rogers' personal and professional fortunes went from good to bad to worse. After the end of the French and Indian War, Rogers married the daughter of a minister, Elizabeth Browne. He tried his hand at several trades including buying land for speculation and fur trading. He even tried to build a road using a lottery to finance it. Debt problems, including his claim against the Crown, led him to London. He had no luck trying to settle these claims.

When he returned to the American colonies the following year he had a new commission, a command at Fort Michilimackinac, and authorization to send out an expedition to discover the Northwest Passage. Within three years Rogers had been relieved of his command, arrested, charged with high treason, and court-martialed, all on trumped-up charges, based mostly on hearsay. His court-martial resulted in an acquittal and he left for London again to attempt to settle his still-not-concluded claim with the Crown and to reclaim his name and honor. In London, Rogers was in and out of debtors' prison, once for almost three years. Finally, after discharging his debts through the new bankruptcy law and being granted a military pension, Rogers again returned to America. Prior to his departure, King George III signed an order forbidding Rogers being given *any* command. Rogers landed in Maryland in August 1775.

He first visited his family in New England and then went to Philadelphia, where he was arrested by the Committee of Safety. The reason given for the arrest was that Rogers was a retired British Army major, on half-pay. During the years that Rogers had been living in England he had paid little attention to developments in America and so he was generally unfamiliar with, perhaps even ignorant of, the independence movement then in full bloom in the colonies. He was equally mystified by his arrest. The arrest was short lived. The Continental Congress intervened and ordered Rogers' release. The release order said, in effect, that if the only reason for the arrest was the fact that Rogers was on half-pay, then the committee should release him, after he gave his word not to make war against the colonies. Rogers gave his parole and was released.

Over the next several months Rogers traveled to various places in the colonies seeking land grants, grants that he would be able to eventually convert to cash to pay off his still hounding creditors. Unbeknownst to Rogers, during this period his movements were closely watched and reports, many of them containing both lies and suspicion, were sent to General George Washington. Rogers was "strictly examined" by several individuals, including army officers, about his activities and intentions. In early February 1776 Rogers, in a meeting with British General Henry Clinton, was asked if he would give services to the Crown. It seems the king had changed his mind. Rogers declined, reminding Clinton of his parole to the Americans. Additionally, almost all of Rogers' comings and goings were reported, not always factually, in local newspapers. Washington received the various reports and was still not certain where Rogers' loyalties were.

In late March 1776 Rogers, still seeking land grants in vain, decided to seek a commission from the Americans. Over time, word of Rogers' dealings with the Americans leaked out and various British reports to London told that "Major Rogers commands a powerful force of Indians" and other untruths. Now each side suspected Rogers of dealing with the other. At this point a twist of fate intervened.

After obtaining several recommendations in New Hampshire to support his request for an American commission, Rogers headed south to collect some others before presenting them to Washington. In New Hampshire a Tory plot was uncovered of unbelievable proportions (and what should have been unbelievable details) that included murdering General Washington and burning New York City. Rogers was featured prominently in this "plot." On 2 July, two days before the Declaration of Independence was signed, the New Hampshire House of Representatives ordered Rogers' arrest. Rogers was apprehended in South Amboy and taken to Washington, who interrogated him at length. Rogers was completely forthcoming with General Washington, including his secret offer of services to the Americans.

Washington read the letters of recommendation but told Congress that Rogers was "not to be sufficiently relied on" and recommended that Rogers' offer, even if genuine, should be refused. Rogers was held in jail. Deciding that the American arrest negated any obligation imposed by his parole, Rogers escaped from jail on the night of 8 July and, ten days later, presented himself to General William Howe in New York and offered his services to the British. On 6 August, Howe authorized Major Rogers "to raise a battalion of Rangers." Although this battalion was named the Queen's American Rangers or, more commonly, the Queen's Rangers, the American newspapers quickly took to referring to them as Rogers' Rangers.

Rogers lost no time in his recruiting drive. For the most part he selected men he knew who had previous military experience and commissioned them as company commanders (captains), authorizing them to recruit companies, usually 50 or so men. Some of his commanders were commissioned based on their claim to be able to recruit a company. The resulting battalion comprised men who were nowhere near the standards of the Rangers from the French and Indian War. Of course, the British were not the only ones recruiting—and it was harder to find loyalists in the colonies. What Rogers ended up with were "farmers and townspeople who scarcely knew one end of a gun from another" and "few had any experience as soldiers."

Headquarters for Rogers' new Ranger unit was on Long Island, New York, near Flushing. At first, hardly anything was known of Rogers' activated commission or of his new unit. In late August 1776, William Lounsbury was caught with a number of recruits for the Rangers. He had enlisting orders from Rogers and a list of men he had recruited when he was captured. Soon after his capture, Lounsbury was bayoneted to death. Reports of Lounsbury's capture and death provided Washington with the first information confirming that Rogers was, in fact, now an active British officer. Washington immediately publicized what he knew about Rogers to spark recruitment for his own forces.

In September, while still on Long Island, an informant told Rogers of the arrival of two men thought to be American spies, Captain Nathan Hale and Sergeant Stephen Hempstead. Rogers began looking for them. Unknown to Rogers, Hempstead had already returned to American lines. Rogers caught up with Hale several days after the American had landed. Rogers eventually convinced Hale that they were on similar missions. The next day Hale carelessly told Rogers that he was on a spying mission for Washington. Rogers immediately arrested Hale and turned him over to General Howe at Howe's headquarters in Manhattan. Hale was hanged the following day, after he made his final statement on the scaffold.

An active correspondence between Washington and Jonathan Trumbull, governor of Connecticut, included a discussion of the recruiting efforts of Rogers and his captains, in an exchange of letters dated 30 September, 4 October, and 13 October 1776. The last letter discussed intelligence about a planned raid by Rogers "into the towns of Greenville, Stamford, and Norwalk …" Whether the plans for such a raid were real or not seemed not to matter; Rogers' name was enough to terrify the countryside surrounding those villages.

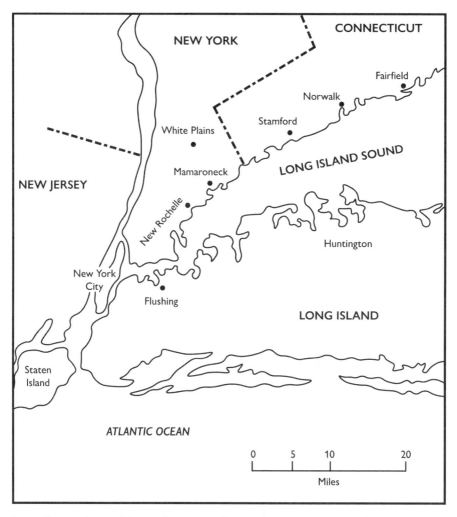

Area of operations of Rogers' Rangers during the American Revolution. (Map by author, digitized by Casemate)

In mid-October, the British Army moved and Rogers' Rangers moved with it, fitting into its right wing in the vicinity of New Rochelle. Elements of Captain John Eagles' company were the first Rangers to enter New Rochelle. On 20 October, the Rangers were ordered to capture Mamaroneck. The Rangers encountered little opposition from local militia units in the operation, capturing or destroying supplies that had been left there for the American forces. These supplies included "rum, molasses, flour and pork." Rogers used a local school for his headquarters and set up his outposts—to the north and east of the village. Since the British Army was to his west, Rogers set up only a weak outpost in that direction. The Rangers' strength at this point is not certain, and it would rarely have exceeded 500 anyway; at least 72 Rangers were on detached duty with artillery units and another 120 were with engineer units.

Sometime after Rogers had established his outposts around Mamaroneck, a one-man reconnaissance was conducted of the British disposition by Rufus Putnam, an American spy, and the information was passed to the American command. A composite force that included soldiers from Virginia, Maryland, and Delaware was formed under the command of an excellent field officer, Colonel John Haslet. Haslet was ordered to attack Rogers' position with his force of 750 men. Using men from the local area to guide his unit, Haslet swung his unit to approach Mamaroneck from the southwest. In the dark the Americans approached a one-man outpost, manned by an American Indian, and caught the sentry by surprise, quickly killing him before he could raise an alarm. Just prior to this attack, Rogers had surveyed his position and the disposition of his outposts, and decided that he was vulnerable in the west. Rogers sent for Captain Eagles and ordered him to take his company to secure the western approach.

Eagles and his 60 men had just settled in when they were attacked by the Americans, who believed they had found the main body of the Rangers. Eagles immediately sent Rangers back to warn Rogers of the contact and continued to direct his unit's defense. But the attack was overwhelming. The Rangers were in danger of being overrun. Some Rangers surrendered; the Americans later reported the names of 36 prisoners. Some Rangers adopted a ruse that worked very well in the dark—in the midst of almost hand-to-hand fighting they began shouting curses at Rogers and the other Rangers as they withdrew. This tactic fooled the American raiders and pulled them closer to the main body of Rangers as they chased the withdrawing Rangers. An element of Virginia soldiers, led by a Major Green, had moved so easily through Eagles' company that a runner reported to Haslet that they had pushed all the way through the main body.

Rogers was notified of the attack in time to have his men standing to when the Americans approached. In the dark the approaching Virginians heard Rogers order his men, "Steady, boys, steady! Fire! Fire!" The onslaught caught the Americans completely by surprise. Not knowing who was opposing him, Haslet ordered his unit to break contact immediately.

Despite his ignorance as to who opposed him at the battle, Haslet reported, and newspapers enthusiastically repeated the lie, that Rogers had turned tail and "skulked off in the dark …" One paper charged: "This blow will ruin the Major's Rangers." It did nothing of the kind. The newspaper stories did, however, fuel the battle between Rogers and his officers on one side and regular British officers on the other. The regulars held the Ranger officers in low regard since they, unlike the British "gentlemen," had not purchased their commissions but had earned them from Rogers based on his assessment of their worth. Purchasing was a common practice in the British Army at that time and for almost the next 90 to 100 years. At any rate, General Howe was well aware of the truth of the encounter near Mamaroneck.

Later in October 1776, Rogers and elements of his Rangers conducted a patrol in Bedford. This patrol returned with several naval prisoners, captured en route, and with a 120-man company of Ranger recruits which they met on the way back. Recruiting for the Rangers continued.

In mid-November, Washington sent General Charles Lee to attack a location east of New York where Rogers' Rangers were camped. Lee's force, which included General John Glover's Marblehead Mariners, attacked but was not successful. In mid-January 1777, elements of the Rangers were located at Fort Knyphausen when it was attacked by an

American force. When the British rejected the American demand to surrender, the Americans withdrew.

Also in January, a new Inspector General of Provincial Forces was appointed by the British Army to review loyalist units, including the Rangers. Alexander Innes, the newly appointed officer, apparently agreed with the sentiments of the regular officers who looked down their noses at Rogers and his Rangers. Citing, incorrectly in places, the backgrounds of some of the Ranger officers and the fact that the ranks included "Negroes, Indians, Mullattos, Sailors, and Rebel Prisoners," Innes submitted his report to General Howe. On 15 February 1777, Rogers, by now a lieutenant colonel, was relieved of command. Between that date and 15 October 1777, the Rangers were commanded by two regular officers before Colonel James G. Simcoe, a regular officer who understood Ranger tactics, became the commander of the Queen's Rangers. Simcoe led the unit until the end of the war.

As for Rogers, his fortunes continued to fail. For a while he continued to recruit for the British, both in America and in Canada. He went back to London to seek some land grants there but returned without any. Once again in America, Rogers was authorized, on 1 May 1779, to recruit two battalions to be known as the King's Rangers. Rogers appointed his brother James as his second-in-command and began recruiting. His efforts did not meet his expectations and by August of that year he gave up his attempts to form a new Ranger unit.

Debt haunted Rogers for the rest of his life. Finally, on 18 May 1795, broke and sick, he died in Southwark, England. His funeral was held in the rain and attended by two unidentified mourners.

Sources

An earlier version of this chapter appeared on Military History Online: https://www.militaryhistoryonline.com/Century18th/RogersRangers.

Books

Cuneo, John R. *Robert Rogers of the Rangers*. New York: Oxford University Press, 1959.
Rose, Alexander. *Washington's Spies: The Story of America's First Spy Ring*. New York: Bantam Books, 2006.

Articles

Cuneo, John R. "The Early Days of the Queen's Rangers: August 1776 to February 1777." *Military Affairs* (Summer 1958): 65–74.

Hutson, James. "Nathan Hale Revisited." *Library of Congress Information Bulletin* (July–August 2003): 168–172. https://www.loc.gov/loc/lcib/0307-8/hale.html.

CHAPTER 2

The Andrews Raid

In early April 1862, Union forces in Nashville, Tennessee, sought to isolate the town of Chattanooga. This was to be one of the early steps in an attempt to push Confederate units in Tennessee south and east, thus shrinking the size of the Confederacy in that area. One of the immediate targets of the overall plan was the railroad junction in Chattanooga. This was a major connection to Memphis to the west and to Atlanta to the southeast.

A plan was concocted at the headquarters of Major General Ormsby Mitchel, which was located in Nashville. The plan called for the capture of a train bound from Atlanta for Chattanooga and, using the train as a means of travel, to destroy intervening bridges, railroad tracks, track switches, and telegraph lines between the two towns, thus breaking a valuable supply line to Chattanooga.

On the evening of Saturday, 5 April 1862, the plan was briefed to James J. Andrews in Shelbyville, Tennessee. Andrews was a civilian but had been one of the most valuable spies in the Union forces for just less than a year. He was a native of Tennessee and had previously worked as a music teacher, a house painter, and as a smuggler. He had operated principally in Georgia, posing as a Confederate loyalist, and was chosen to lead the raid because of his knowledge of the area.

Over 6 and 7 April Andrews interviewed 30 soldiers from three Ohio infantry regiments who had volunteered to take part in the raid. Two of these men, Private William Knight of the 21st Ohio, who had previous experience as a railroad engineer, and Private Charles Shadrack of the

2nd Ohio, who had been a mechanic, were immediate selections for the raiding team. Andrews selected another civilian, William Campbell, and a total of 22 soldiers—five from the 2nd Ohio, nine from the 21st Ohio, and eight from the 33rd Ohio—to accompany him on the raid.

Once they were selected for the raid, Andrews told the soldiers to arrange to borrow or buy civilian clothes, for that is what they would wear on the operation. On Monday night, 7 April, Andrews met with his raiding team on the outskirts of town. He briefed them on the plan and instructed them to travel to Marietta, Georgia, in small groups over the next three days. Their cover was that they were Confederate sympathizers traveling to Atlanta to enlist in the Confederate army. They were to board the 5 p.m. train in Marietta (bound for Chattanooga) on Thursday, 10 April. It rained heavily during the intervening days, slowing travel down considerably. Two of the raiders (Private James Smith of the 2nd Ohio and Corporal Samuel Llewellyn of the 33rd Ohio) were detained by Confederate forces in Tennessee and were forced to enlist on the spot. The others did not make it to Marietta until Friday. This delayed the raid until the next day.

On Saturday morning, Andrews went over the final details of the plan with 17 of his raiders in his hotel room. They all left immediately to board the train for Chattanooga. Unknown by the others, two of the raiders (Private John Porter of the 21st Ohio and Corporal Martin Hawkins of the 33rd Ohio) overslept because they had not been called at their hotel in time to catch the train.

With 18 of the raiders on board, the train, which was pulled by the locomotive *General*, left the Marietta station. Shortly after, at 6 o'clock, the train pulled into the small town of Big Shanty for a 20 minute stop to take on additional fuel and water. While the train was being serviced in preparation for the sometimes hilly and steep journey to Chattanooga, the passengers disembarked to eat breakfast at the nearby Lacy Hotel. Big Shanty was chosen as the place where the raiders would hijack the train because there was no telegraph station there. On one side of the train was the hotel; on the other side was an encampment of Confederate soldiers from Camp McDonald.

When the passengers had all disembarked and entered the hotel, Andrews and Knight checked the engine and uncoupled all but the tender and three box cars from the engine. Knight started the engine and the remaining raiders boarded the train. The train pulled out, completing the first step in the raid.

Left behind in Big Shanty were the passengers and crew of the train. One of those eating breakfast was William Fuller, the *General's* conductor. As soon as he saw the train pulling out of the station, Fuller grabbed two of the crewmembers, a foreman and an engineer, and they began running after the train. Fuller thought that the thieves were deserters from the Confederate camp across from the hotel. This led him to believe that they would likely use the train for a short distance and then continue on foot. He also was aware of the terrain to the north of Big Shanty and that the train would probably only be able to average about 15 miles an hour during the trip. He believed that he would be able to commandeer another train to give chase at some point, which was why he gave chase in the first place.

Fuller was right about one thing. The train *did* stop just north of Big Shanty so the raiders could cut the nearby telegraph lines. While doing this they noticed a pile of railroad ties near the track and piled some of them across the tracks to cut off any train that might come after them and put the remaining ties in one of the boxcars.

At this stop Andrews told his men that once they had passed only one train the way would be clear for them to go straight through to Chattanooga, burning bridges and cutting the line as they went. The train started up again and moved north of Altoona, the next town to the north. When it stopped again, this time the raiders began to take out a rail from the tracks. This took longer than Andrews thought it would. As soon as the rail was freed, he told his men to store it in one of the box cars with the others.

As they rode past the next bridge, Andrews noticed a small locomotive, the *Yonah*, parked in a siding. Fuller and the crewmembers continued running. By this time Fuller was really certain that he was not chasing deserters, but "Yankee spies."

By now Andrews had worked out a story to tell at the various stations the raiders passed through. He would tell the waiting passengers and railroad workers that they were on a priority mission to deliver weapons

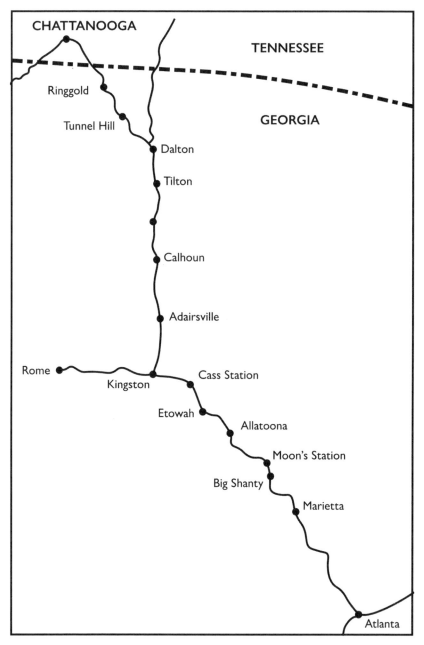

Sketch map showing the anticipated extent of the Andrews raid. (Map by author, digitized by Casemate)

and ammunition to Confederate General P. G. T. Beauregard. To back up his story, Andrews had mounted a red flag on the train, which signaled his emergency priority and also let waiting passengers know that another train was following at a distance. At one of the early stations, one of the raiders borrowed a crowbar from a rail worker. The raiders had neglected to bring this valuable tool with them for fear of drawing undue attention to them as they made their way south. It was to pry rails loose.

When the raiders reached Cass Station, they stopped to take on water and additional wood. They worked quickly and left soon after they had stopped. At the next station, Kingston, the raiders were ordered onto a siding. They were delayed by a priority ammunition train heading *south*. The delay lasted an "agonizing" 35 minutes for the first section of the train to pass through. It was another 10 minutes before they could leave, delayed further by the second section of the priority train. Andrews was afraid that the railroad workers at the station were becoming suspicious of his train.

By this time Fuller and his crewmembers had found a handcart, piled on, and began pumping. They passed through Altoona and continued north. Almost immediately Fuller spotted the *Yonah*. Just as he did the handcart began a downhill run, moving pretty quickly. They were watching the *Yonah* and didn't see the gap in the railing until it was too late. The handcart flipped over and tossed the riders. However, Fuller was not injured. He got back to his feet and ran to the locomotive. He got it started and continued after the raiders.

After another 10 miles, the raiders stopped again to remove another rail. While they worked to take out the rail, the raiders heard the noise of the *Yonah* in pursuit of them. One of the raiders, Private Knight, the engineer, thought to himself (as he related later), "… the destruction of a bridge was the only way that would save us and to do this we had to outrun them."

When he got to Kingston, Fuller abandoned the *Yonah* and commandeered another, bigger locomotive, the *William B. Smith*. He also recruited several more railroad employees to assist him in the pursuit. He took up the chase again. As the pursuers came upon obstructions on the tracks, all on board jumped off and worked to clear them. When they reached the gap in the tracks, Fuller abandoned the *William B. Smith* and took off on foot again.

When the raiders reached Adairsville they were told they would have to delay for another southbound train that was already on its way. Andrews shook off this delay and left the station at once, ordering full speed once they had cleared the station. As it turned out, Andrews' gamble worked to his advantage. When the raiders arrived in Calhoun the southbound train was stopped in a siding. Andrews stopped long enough to talk to the conductor of the southbound train, asking why it had a red flag on the rear car (as Andrews knew, signaling that another train was coming behind it). The conductor explained that Union General Mitchel had launched an attack on Chattanooga from Stevens, Alabama. The Confederate military authority there ordered as many trains as possible to move south to evacuate essential military materials out of the city and to avoid possible capture by Mitchel. Andrews realized that he had no "authority of movement" over the southbound trains and would have to give way to any he encountered. Any delays thus caused would give Fuller time to close the distance between the two trains. Andrews immediately ordered the *General* out of the station, running again at full speed.

Before he reached Adairsville Fuller encountered a southbound train, pulled by the *Texas*. He managed to stop this train and explained to the conductor what he was doing. The crew of the *Texas* agreed to let Fuller take control of the train in pursuit of the raiders. Fuller continued on his way, running the *Texas* in reverse. When he arrived in Adairsville, Fuller shunted all of the cars except the locomotive onto a siding. When this was completed, he continued the chase north, still running the *Texas* in reverse, picking up speed as he went and blowing the train's whistle. This was to warn other trains to get out of his way. He told them to take the first siding they came to and remain there until the *Texas* passed them by. As he arrived in Calhoun, Fuller slowed down and stopped. He persuaded 11 Confederate soldiers and a telegraph clerk at the station to get on the *Texas* to help in the chase. Then he took off again, still running in reverse and blowing his whistle.

As Andrews headed north of Calhoun he began to calculate what still needed to be done. Between cutting telegraph lines, burning several bridges, and taking out rails, he came to the conclusion that the raiders would need to make about seven stops, including one to take on more wood and water. The first stop was at a place where the tracks curved. While the raiders were working on the rail, they heard the whistle of the *Texas*. Although they had only been able to bend one rail loose and lift it part-way up, they broke off, boarded the *General*, and took off again.

When Fuller came to that curve the *Texas* was moving very fast. Fuller didn't see the loosened rail until he was almost on it. However, he noticed that the bent rail was on the inside of the curve. He realized that because the *Texas* was travelling fast its weight would be thrown to the outside of the track. All of this happened in an instant, too late for Fuller to have done anything even if he thought he could. When the *Texas* hit the rail, those on board felt a jolt but the train continued running.

Looking back, Andrews could see that the *Texas* was still pursuing him and that the bent rail had not really slowed it down much. He made a quick decision and told the raiders to unhook the last boxcar. This would, after a short while for it to slow down, make it reverse course and careen back along the tracks into the *Texas*.

When the boxcar came into sight, Fuller slowed the *Texas* down so that he could hook up the boxcar to his locomotive. Once it was connected, he picked up some speed. On board the *General*, Andrews had his raiders begin to throw railroad ties onto the tracks, hoping to slow the *Texas* down. This worked. Each time the *Texas* approached another tie Fuller had to slow it down so that those on board could get off and clear the tracks. The *General* began pulling away, increasing the distance between it and the *Texas*.

When he eventually ran out of railroad ties, Andrews ordered a second boxcar cut loose. Fuller again had to slow down even more to get it attached, just like the previous boxcar. This worked to slow down Fuller for a while but it didn't last long. As he came to the station in Resaca, Fuller let one of his crew get off the train to reset the switch to a siding. Then he cut the boxcars loose and pushed them down the siding. After resetting the switch, the *Texas* continued the pursuit on the main track, still running in reverse.

Andrews had just enough of a lead that he was able to stop in Tilton to take on more water. However, the wood there was soaked from all the rain and was not useable. Because Fuller was still coming on Andrews would no longer afford the time to stop to tear up tracks or set bridges on fire. He had also left the telegraph line at Tilton intact. He now had to concentrate what time he had to escape northward.

When Fuller got to Tilton he noticed the uncut telegraph line. He dropped off the telegraph clerk, telling him to get the word to Chattanooga about the stolen *General*. When this message was received at the garrison in Chattanooga, troopers were sent south with orders to capture the raiders. Andrews was now being pursued from both north and south although he was not aware of it. To make matters worse, it had now begun to rain.

The chase continued on through the stations of Dalton and Tunnel Hill. South of Ringgold the tracks led through a tunnel. Andrews figured he would gain some time here, suspecting that Fuller would slow down, anticipating an ambush. The raiders tried to set fire to the last boxcar but they didn't have enough time and the wood was too damp. After three attempts, Andrews realized that he was out of time. He ordered his raiders off the train and told them to take off, heading west toward what he hoped were Mitchel's lines. He also sent the *General* back down the tracks in reverse. Then he too fled. His last effort, however, failed to stop Fuller, who was able to meet the *General* before it had had enough time to pick up much speed. The chase on the railroad tracks was now over. It had run for 87 miles and had taken just over seven hours. Andrews had gotten within 18 miles of Chattanooga before he and his men had escaped.

Over the course of the next two weeks, Confederate troops arrested all of the raiders, including the two who had been left behind in Marietta. Andrews and seven of the prisoners, including Campbell, were tried in a military court as spies. All of them were hanged in Atlanta. The others were sent to a military prison without a trial. In October 1862, eight of the prisoners escaped and were able to reach Union lines. The remaining six were eventually paroled in March 1863.

On 25 March 1863, the paroled six were in Washington, D.C. There the Secretary of War, Edwin Stanton, presented them with the Medal of Honor, which had recently been approved by Congress. They were the first to receive this award, which was awarded first to Private Jacob Parrott of the 33rd Ohio. Eventually those who had escaped and reached friendly lines also were awarded the Medal of Honor. Because they were civilians, Andrews and Campbell were not eligible. In time Congress approved the award for the others who had been hanged.

The entire list of those who took part in the Andrews raid is as follows:

> Civilians: James J. Andrews, William H. Campbell.
>
> 2nd Ohio: Sergeant Marion A. Ross, Corporal William Pittenger, Private Charles P. Shadrack, Private James Smith, and Private George D. Wilson.
>
> 21st Ohio: Sergeant Elihu H. Mason, Sergeant John M. Scott, Private William Bensinger, Private Wilson W. Brown, Private Robert Buffum, Private William J. Knight, Private John R. Porter, Private John A. Wilson, and Private Mark Wood.
>
> 33rd Ohio: Corporal Daniel A. Dorsey, Corporal Martin J. Hawkins, Corporal Samuel Llewellyn, Corporal William H. H. Reddick, Private Jacob Parrott, Private Samuel Robertson, Private Samuel Slavens, and Private John Wollam.

Sources

Books

Aiken, Gene, ed. *The Great Locomotive Chase: As Told by Men Who Made it Happen.* Gatlinburg, TN: Historic Press, 1994.

Angle, Craig. *The Great Locomotive Chase: More on the Andrews Raid and the First Medal of Honor.* Mercersburg, PA: Mercersburg Printing, 1992.

Epstein, Samuel, and Beryl Epstein. *The Andrews Raid or The Great Locomotive Chase—April 12, 1862.* New York: Coward-McCann, 1956.

Pittenger, William. *Daring and Suffering: A History of The Great Railroad Adventure.* New York: Time-Life Books (reprint), 1982.

Roberts, MacLennan. *The Great Locomotive Chase.* New York: Dell Books, 1956.

Zedric, Lance Q., and Michael F. Dilley. *Elite Warriors: 300 Years of America's Best Fighting Troops.* Ventura, CA: Pathfinder Publishing, 1996.

Articles

Adwar, Corey. "The Bold Civil War Raid That Led to the First-Ever Medal of Honor," Task & Purpose. Published April 13, 2015, https://taskandpurpose.com/military-life/the-bold-civil-war-raid-that-led-to-the-first-ever-medal-of-honor/.

CHAPTER 3

American Airborne Units in World War II

This short history will cover those combat military groups, squadrons, battalions, regiments, divisions, and the one corps of U.S. airborne units in World War II. It will not include Army Air Corps units (such as the Air Commandos) or Troop Carrier units, or organizations that had American individuals in them who were airborne qualified and even made operational jumps, such as Army and Marine Corps members of the Office of Strategic Services (including those with Jedburgh teams and Operations Groups), or the multi-service, multi-national Special Allied Airborne Reconnaissance Force (which initially included women who had previously jumped into denied areas).

Background

The first plan to use parachute forces by American units was developed during World War I. On 17 October 1918, Brigadier General William P. (Billy) Mitchell, a later proponent of strategic aerial bombing, conceived the idea of dropping an American division by parachute from bomber aircraft into an area in the vicinity of Metz, Germany. The details of the planning were developed by Major Lewis H. Brereton, a member of Mitchell's Air Service staff. Brereton would later serve as the commander of the First Allied Airborne Army during World War II as a lieutenant general. (The First Allied Airborne Army consisted of the American

XVIII Airborne Corps, which included the 17th Airborne Division, the 82nd Airborne Division, and the 101st Airborne Division; and the British I Airborne Corps, which included the 1st Airborne Division and the 6th Airborne Division. Troop carrier units were also part of Brereton's command.)

In the event, the plan for the Metz operation, which would have air-dropped the U.S. 1st Infantry Division, was cancelled because of the Armistice on 11 November 1918 that ended the war. In the intervening years between the world wars, the U.S. military establishment paid little attention to the idea of airborne warfare. Russia and Germany, however, developed forces within their respective militaries that relied on aerial delivery of troops, including by parachute and glider. The U.S. Army conducted brief experiments in 1928 with parachute forces but discontinued them shortly after. It was not until 1938 that serious consideration of airborne warfare was again raised, this time at the Command and General Staff College at Fort Leavenworth, Kansas. This effort was in the form of exercises that involved the formation of doctrine for the employment of airborne forces.

U.S. Army

The formation of American airborne units began on 26 June 1940, with the establishment of the Parachute Test Platoon (PTP) at Fort Benning, Georgia, with volunteer members from the 29th Infantry Regiment. This unit was formed to test the feasibility of parachute operations in the U.S. Army. The PTP began training immediately and made its first jump on the morning of 16 August 1940. Although early military planners thought that paratroopers would most likely operate in small units against "communications and supply installation in enemy rear areas," this concept soon changed. Eventually several Parachute Infantry Battalions (PIBs) were activated at Camp Mackall, North Carolina, Camp Toccoa, Georgia, and Fort Benning. They were manned by graduates of the Airborne School. These battalions included the 501st, 502nd, 503rd, and 504th, as well as others. As decisions were made to create airborne divisions, these battalions were increased in strength to become Parachute Infantry Regiments (PIRs), generally with the same numerical designation. In a history that is not necessary to detail here, the 2nd Battalion, 503rd PIR was eventually redesignated as the 509th PIB and will be referred to that way hereafter.

The first group of glider pilots completed their training in June 1942. The first Glider Infantry Battalions (GIBs) were created on 5 September 1942 and soon after were upgraded to Glider Infantry Regiments (GIRs). By this time, decisions had been made to man the several airborne divisions with two or three PIRs and one or two GIRs, although initially the preference had been for two GIRs and one PIR. Soldiers in glider units, unlike the paratroopers, did not receive hazardous duty pay.

On 15 August 1942, the 82nd Infantry Division and the 101st Infantry Division were redesignated as airborne divisions at Camp Claiborne, Louisiana. The 82nd was assigned the following regiments: the 504th and 505th PIRs and the 325th GIR, along with parachute and glider field artillery, engineer, and other units (signal, military police, medical, intelligence, etc.). The 507th and 508th PIRs, originally deployed to England as the 2nd Airborne Brigade, were added to the 82nd when it deployed to England following operations in Sicily and Italy. The 507th and 508th were added because the 504th PIR had remained in Italy. Following the D-Day operations, the 507th PIR was transferred to the 17th Airborne Division and the 504th was brought back in to the division from Italy. The 82nd conducted combat jumps (two separate jumps on succeeding nights) in Italy, and combat jumps and glider operations in Sicily, France, and The Netherlands.

The 101st was initially assigned the following regiments: the 502nd PIR and the 327th and 401st GIRs along with the same mix of additional units as the 82nd. Eventually, the 501st and 506th PIRs were added to the division and the soldiers of the 401st GIR were split up between the 325th GIR (of the 82nd) and the 327th GIR. This led to the disbanding of the 401st. The 101st conducted combat jumps and glider operations in France and The Netherlands.

The 509th PIB was always an independent unit throughout the war. It was deployed to England, and eventually made the first combat jumps by American airborne units in North Africa. It later operated in Italy and southern France, where it also conducted combat jumps. The 503rd PIR remained an independent unit throughout the war and was deployed to the Pacific area of operations. The 503rd conducted combat jumps in New Guinea and the Philippines.

The 11th Airborne Division was activated on 25 February 1943 at Camp Mackall and contained the 511th PIR and the 187th and 188th GIRs, along with the same mix of additional units as the other airborne divisions. The 11th was deployed to the Pacific area of operations where its soldiers were cross-trained in parachute and glider operations. The 11th conducted combat jumps and glider operations in the Philippines.

The 17th Airborne Division was activated on 15 April 1943 at Camp Mackall and consisted of the 513th PIR and the 193rd and 194th GIRs, along with the same mix of additional units as the other airborne divisions. It deployed to England and then Europe (adding the 507th from the 82nd), where it fought during the war. The 17th conducted a combat jump and glider operation in Germany.

The 82nd, 101st, and 17th were incorporated into the XVIII Airborne Corps, which had been redesignated from the XVIII Corps on 27 August 1944. The XVIII Airborne Corps conducted combat jumps and glider operations in the Netherlands and Germany.

The 13th Airborne Division was activated on 13 August 1943 at Camp Mackall and included the 515th PIR and the 88th and 326th GIRs along with the same mix of additional units as the other airborne divisions. Eventually, following its deployment to England, the previously independent 517th PIR, which had made a combat jump and seen action in southern France, was added to the division. The 13th was kept in strategic reserve throughout the war, ultimately deploying to France in early 1945, but never was deployed in combat.

In addition to the 517th PIR and the 509th PIB, two independent battalions which had been formed in the Canal Zone (the 551st PIB and the 550th Airborne Infantry Battalion, a glider unit), operated under the auspices of the First Airborne Task Force commanded by Major General Robert T. Frederick, the former commander of the First Special Service Force (see below). The First Airborne Task Force was a composite U.S., British, and French corps-sized unit that conducted operations in southern France. All of these units conducted a combat jump and glider operation in southern France. The First Airborne Task Force was disbanded in November 1944.

The 541st and 542nd PIRs, formed from the 1st Airborne Training Battalion, conducted parachute training of soldiers at the Airborne School at Fort Benning and acted as a holding unit for soldiers who rotated through them to units overseas. Both units moved between Fort Benning and Camp Mackall. The 541st PIR was activated on 12 August 1943 at Fort Benning. In July 1945, the 541st was deployed to the Philippines and assigned to the 11th Airborne Division, pending operations directed against the Japanese home islands. Soon after arriving, the men of the 541st were reassigned to the other regiments in the 11th Airborne Division and the regiment was deactivated, much to the disappointment of its men.

The 542nd PIR was activated on 1 September 1943, also at Fort Benning. On 17 March 1944, the 542nd was deactivated and reactivated as the 542nd PIB, moving to Camp Mackall on 1 July 1944. At Camp Mackall, the 542nd tested new techniques and equipment for the Airborne Center Command Headquarters. On 1 July 1945, the 542nd was reflagged as the Airborne Center Training Detachment. The 503rd Parachute Infantry Regiment was formed in February 1942. It took units from the 503rd and 504th battalions. When it reached the Pacific Theater a battalion that had been training in Panama was added to the regiment. The 503rd conducted jumps in New Guinea, on Noemfoor island, and on Corregidor island.

The 555th PIB, the only all-black airborne unit in the U.S. Army, was activated initially as a company on 25 February 1944 and was upgraded to a battalion on 25 November 1944. Although the 555th never left the United States, it was given a classified mission to locate, disarm, and return as much of the equipment as possible for intelligence exploitation from Japanese balloon bombs. The baskets on these balloons contained incendiary devices on them. The balloons were being sent by the Japanese, in retaliation for the Doolittle Raid in 1942, along the prevailing winds to crash and start fires in North America. The 555th operated under the cover of military Smokejumpers with the U.S. Forest Service in California and the Pacific Northwest. The 555th also assisted in training pilots of the U.S. Navy in low level bombing operations in support of ground troops. All told, the 555th conducted over 1,200 individual jumps to fight 36 forest fires.

The various U.S. theater commands created so-called "ghost divisions" as part of several deception programs to make the enemy believe that more units were available than there actually were. This included five ghost airborne divisions.

Included in the things they did to reinforce the existence of these ghost units were such things as: having soldiers wear the shoulder patches on their uniforms, having items published in local newspapers that discussed the units and even included photographs, and used rubber items such as tents and vehicles (including tanks) that looked real from a distance both at ground level and from an airplane's height.

The First Special Service Force was the only military unit in the U.S. Army that comprised both U.S. and Canadian soldiers. Formed in response to plans to use a mechanized sled type vehicle in the snow in Norway, the Force was activated on 9 July 1942 at Fort William Henry Harrison in Helena, Montana. Force men received training, with emphasis on night-time execution, in techniques of parachute operations, snow and mountain operations, and amphibious warfare. Their training also included the use of explosives for demolition and a new method of hand-to-hand fighting, known as the O'Neill System. Eventually, the plan to use the snow sled in combat was dropped. However, the Force continued to train for combat. The Force first deployed to Kiska, Alaska, in the Aleutian Islands, where the operations plan called for one of its regiments to jump behind Japanese lines; this jump was later cancelled and the entire Force was committed as a conventional infantry unit in the fighting, making an amphibious landing. The First Special Service Force fought in Alaska, Italy, and as part of the First Airborne Task Force in southern France. The First Special Service Force was disbanded in December 1944 and the Americans from the unit, along with members of the 99th Infantry Battalion and the survivors of Darby's Rangers (the 1st, 3rd, and 4th Ranger battalions), were incorporated into the 474th Infantry Regiment.

U.S. Marine Corps

The U.S. Marine Corps also became interested in parachute units. In October 1940, the 1st Marine Parachute Battalion (MPB) began its airborne training at Naval Air Station Lakehurst, New Jersey. The 2nd MPB trained there as well, in December 1940. The 3rd MPB received its training at Camp Kearney, California, near San Diego. Soon after, a parachute training center was opened at Camp Elliot, next to Camp Kearney, and another training center was established at New River, North Carolina. These three MPBs were deployed separately to the Pacific and fought there in conventional infantry roles, including conducting some amphibious landings. Eventually, on 1 April 1943, the 1st Marine Parachute Regiment (MPR) was activated at Vella Lavella, in the New Georgia Group, and included all three of the MPBs. On 2 April 1943, the 4th MPB was activated and remained in a training status at Camp Elliot and later moved to Camp Pendleton, California, where it was disbanded on 19 January 1944. Although the first three MPBs were deployed in combat operations, none of them conducted any operational jumps. A jump was scheduled onto the Japanese airfields of Kahili and Kara on Bougainville but the American planners cancelled the jump, fearing heavy casualties. Units of the 1st MPR fought on the islands of Guadalcanal, Gavutu, and Tanambogo in the Solomon Islands; Bougainville; the island of Choiseul; and in other Pacific battles. All three of the MPRs were returned to the United States by early 1944 and disbanded (along with the Marine Raider units) on 29 February 1944.

The development of the Marine glider program proceeded separately from its parachute program because Headquarters, Marine Corps, did not envision gliders to operate with parachute units. In July 1941, the Marine Corps announced its intention to train 50 officers and 100 non-commissioned officers as glider pilots. Planning was also initiated to procure 75 12-man gliders to transport one "Air Infantry Battalion." The design, development, and procurement of gliders became a difficult course and included many twists, turns, and changes of policy along the way. Glider pilot training eventually began in November 1941 at a civilian training course.

The first gliders were delivered in mid-December 1941 and additional training began at Paige Field, Parris Island, South Carolina. At the same time, the Glider Detachment was created at Cherry Point, North Carolina. In mid-March 1942, Glider Group 71 (MLG-71) was formed to replace the Glider Detachment; it had a proposed organization of 20 officers and 218 enlisted men. The group organization included Headquarters and Service Squadron 71 and Marine Glider Squadron 711 (VML-711), and was initially stationed at Marine Corps Air Station, Parris Island. Glider Group 71 was assigned to Fleet Marine Force for command and control. A more permanent base for Glider Group 71 was later established at Marine Corps Air Station, Eagle Mountain Lake, Texas. Glider Group 71 arrived at the base on 24 November 1942. A second base was established at Edenton, North Carolina, but was never used for gliders. At least two other bases were designated but also were never used for gliders, one at Norman, Oklahoma, and the other at Addison Point, Florida.

Eventually the Marine Corps decided that gliders were not suited to support the planned amphibious operations on the islands in the Pacific; this spelled the end of the glider program. On 24 June 1943, that decision was formalized; the Marine glider program was terminated and Glider Group 71 was disbanded.

Geronimo and the Paratroopers

From almost the earliest formation of airborne units in the U.S. Army, American Indian names or symbols have been used by paratroopers. For example, the 501st PIR used the symbol of an Apache war chief holding a lightning bolt below a chute canopy, with the word "Apaches" contained within the shroud lines of the parachute, superimposed over the name "Geronimo," a famous Chiricahua Apache Indian leader of the American southwest. The 501st unit crest (distinctive insignia) shows an American Indian symbol of a Thunderbird with the unit motto ("Geronimo") on a scroll underneath.

The 506th PIR unit crest (distinctive insignia) features the American Indian name for the mountain near Camp Toccoa, where the regiment trained. Currahee Mountain was the scene of many unit running and other training exercises. Currahee means "stand alone" or, sometimes, "we stand alone." The insignia shows six parachutes descending onto a mountain. The paratroopers of the 506th adopted Currahee as their regimental motto because that was their objective behind enemy lines, to "stand alone."

The First Special Service Force used several American Indian-based ideas in its symbols. Although the First Special Service Force was not on jump status, all of its combat echelon soldiers were jump qualified; a jump was scheduled for one of its regiments in Alaska but was cancelled at the last hour. The unit patch is in the shape of an arrowhead. The branch insignia for both officers and enlisted soldiers incorporated the crossed arrows of the Indian Scouts, which the First Special Service Force used with the written permission of the surviving members of the Indian Scouts. And the soldiers of the First Special Service Force, both Canadian and American, were referred to as "Braves."

In addition to these various names and symbols, the cry "Geronimo!" is associated with early paratroopers as they exited the plane. This cry was later adopted by paratroopers in many units. So, how did this tradition get started? It began with the Parachute Test Platoon at Fort Benning in the summer of 1940. On the night before their first mass jump (they had already completed two individual tap-out jumps), four platoon members had been to the post theater to see the movie *Geronimo*. Later, over several beers, they discussed whether they would be afraid or even aware of their surroundings when they jumped out of the plane the following day. One of them, Private Aubrey Eberhardt, told the others that he would shout the name "Geronimo" when he jumped to prove that he was not afraid. The others all thought this was a good idea and agreed to do the same thing. When the time came, all four of them followed through with their pact from the night before. With their actions that day they originated what became the jumping cry of American paratroopers. The division song of the 11th Airborne Division even contains this phrasing:

> Down from Heaven comes Eleven
> And there's Hell to pay below—
> Shout Geronimo: Geronimo!

Jump Wings

The "silver wings" of the paratrooper were designed by Captain (later Lieutenant General) William P. Yarborough when he was assigned to the 501st PIB. He had been chosen by the adjutant general of the War Department to design and procure a suitable badge to be worn by qualified paratroopers as a symbol of their certification. He was authorized by his commander to choose whatever design he thought was acceptable. Yarborough made the initial design on 3 March 1941 and provided a copy to the army quartermaster general. The approval process of the wings design at the War Department took one week; Yarborough's design was formally approved on 10 March. With the assistance of Mr. A. E. Dubois, of the quartermaster general's office, an order of 350 wings was made to the company of Bailey, Banks & Biddle of Philadelphia. The wings were delivered to the 501st on 14 March. As a final step, to protect the design from any unauthorized reproduction, Yarborough submitted his design to the U.S. Patent and Trademark Office for assignment of a patent. The patent was approved on 2 February 1943. The formal description of the badge is: "An oxidized silver badge 1–13/64 inches in height and 1–1/2 inches in width, consisting of an open parachute on and over a pair of stylized wings displayed and curving inward." In order to be able to wear these wings, personnel must have satisfactorily completed the prescribed proficiency tests while assigned or attached to an airborne unit or to the Airborne Department of the Infantry School, or participated in at least one combat parachute jump.

Glider Badge

A badge for glider soldiers, similar to the paratroopers' jump wings, was eventually designed and approved for issue and wear; official approval came on 2 June 1944. Its formal description is: "An oxidized silver badge 11/16 inch in height and 1–1/2 inches in width consisting of a glider, frontal view, superimposed upon a pair of stylized wings displayed and curving inward." In order to be eligible to wear the glider badge, personnel must have been assigned or attached to a glider or airborne unit or to the Airborne Department of the Infantry School, and satisfactorily completed a course of instruction or participated in at least one combat glider landing into enemy-held territory.

Postscript: Insignia developments after World War II

In the U.S. Army, the advanced qualifications of Senior Parachutist and Master Parachutist (symbolized by a star for Senior and a star and wreath for Master on top of the parachute canopy of the wings) were authorized in 1949.

The Navy and Marine Corps Parachutist Insignia, awarded to members of those services who have been awarded the Basic Parachutist Insignia (Jump wings) and have completed an additional five static-line jumps, was approved for issue and wear in 1963.

Equipment Development

The parachute

The history of the parachute begins long before man learned how to fly. In the late 15th century, Italian genius Leonardo da Vinci imagined what it would be like for man to jump safely from heights. He designed a four-sided pyramid-shaped "tent" from which a man could be suspended as he floated to the ground. Along with his design he made a notation: "If a man has a tent of linen of which the apertures have all been stopped up, and it be 12 *braccia* across [one *braccia* was approximately three feet] and 12 in depth, he will be able to throw himself from any great height without sustaining any injury."

Nothing was done to test da Vinci's theory for about 300 years. And even then it was not tested by a human. In a typically experimental manner, an animal was chosen as the subject. In 1783, in Paris, Jean-Pierre Blanchard, using a parachute of his own design based on da Vinci's description, dropped a dog from a tethered balloon. The design worked perfectly and the dog landed safely. Also in 1783, another Frenchman, Louis-Sebastien Lenormand, jumped from a tree and used two umbrellas fashioned into a kind of parachute to slow his descent. However, there is a school of thought that Andre-Jacques Garnerin, a French magician and balloonist, was the first person to design and test a parachute capable of slowing a man's descent from higher altitude. While there are few details known of this event, it is believed that Garnerin, while a prisoner during the French Revolution, got the idea of using air resistance to slow down a person's fall from altitude. He did not use his idea in order to escape from prison, but continued to work on his idea to use air resistance. Garnerin designed a parachute with a canopy that was 23 feet in diameter and attached to a basket by suspension lines.

It is reported that Garnerin first tested his design on 22 October 1797, by ascending in a balloon to about 3,200 feet, then climbing into the basket attached to the canopy. After cutting the parachute loose from the balloon Garnerin descended in a controlled fall. He did not, however, take into account a method of releasing the air from his parachute. As a result, he swung back and forth wildly in his descent. He did land safely but not near where he had planned. As a result, he altered his design to permit some air to escape through a vent at the top of his parachute. Five years later, Jeanne-Genevieve Garnerin, Andre-Jacques' wife, was the first woman to make a parachute jump, descending from 8,000 feet in England. Garnerin himself was killed in 1823 in a balloon accident while getting ready to demonstrate a newly designed parachute.

In the years following Garnerin's early jumps, daredevil acrobats performed on a bar suspended from beneath a tethered balloon basket, jumping when they had completed their performances. The first recorded death from a parachute drop occurred in July 1837, when Robert Cocking, an Englishman, jumped from "a mammoth balloon." During the descent his parachute came apart. The parachute he used was based on his own design. It resembled an upside-down umbrella and weighed more than 200 pounds because of the metal framework over which the linen parachute was stretched. As a result of Cocking's death, England passed laws banning the use of parachutes.

The next evolution in parachute design and use came in 1884. Thomas and Samuel Baldwin, high wire walkers, drastically redesigned the parachute in a New Orleans restaurant. Until then, early parachutes had consisted of metal or wood frameworks over which linen material was placed. The Baldwins conceived of a parachute that could be packed in a canvas container attached to a harness worn by the jumper. It took three years before the idea could be turned into a practical prototype. In January 1887, Thomas made a successful jump of the collapsible parachute over Golden Gate Park in San Francisco. The parachute itself was positioned on the outside of the balloon basket. Thomas climbed down into the parachute. The weight of Thomas' body caused the lines to unfold and then release the parachute from the pack.

The idea for the parachute to be worn on the body did not come about until after airplane flight was achieved. The first development was a parachute attached to the underside of an airplane, similar in some ways to the Baldwin's idea. This concept was developed by Leo Stevens. The way it worked was that the pilot could jump out of the bottom of the plane into the parachute. The pilot's weight would release the parachute from the container and open as he fell. On 28 February 1912, Albert Berry became the first person to jump from an airplane with a parachute, when he demonstrated Stevens' idea to the U.S. Army at Jefferson Barracks, Missouri. Despite the success of Berry's jump, pilots considered the arrangement to be impractical. Their rationale was that since it would be used strictly as a life-saving piece of equipment, it would take too much time for the pilot to actually release the plane's controls and get into the harness of the parachute for it to be of any use in an emergency.

The next step was the development of a static line parachute. The parachute pack was built in to a coat worn by the user, a pilot. This system was designed by Charles Broadwick. The static line was attached to a strong point on the plane by a hook at the end of the line whose other end was attached to the pack. When the jumper left the plane, the pack static line would pull the pack open and the folded lines inside would play out, pulling the parachute with them. The suspension lines were attached to the skirt of the parachute canopy at one end and to the coat worn by the pilot at the other end. This bulky system proved to be too cumbersome to be used inside the confined space of a plane's cockpit, restricting the pilot's movement.

During the early years of World War I, parachutes were worn mostly by soldiers in observation balloons, which were tethered to the ground.

The parachute was used if planes attacked the balloon and the soldiers had to jump out to save their lives. Eventually, German and English pilots began to wear parachutes and lives were saved when their planes were shot down. The U.S. did not enter the war until mid-1917. Its pilots were not issued parachutes because they were not available. Late in the war an attempt was made to correct this when parachutes were requested for flying personnel.

In a meeting on 17 October 1918 with General John J. Pershing, Colonel William P. "Billy" Mitchell, Pershing's Air Service Officer and chief of American air units, proposed a daring plan. The meeting had begun with a proposal by Mitchell, a proponent of strategic bombing (then and later), to send American bombers into German rear areas. Pershing told Mitchell that his first priority was to achieve air superiority in the immediate battle area by driving German air forces out of the battle zone. His next priority was to attack German ground forces capable of threatening American forces.

Following this discussion, Mitchell proposed a plan to be implemented in the spring of 1919. The plan called for a fleet of bomber aircraft to carry the 1st Infantry Division and drop them by parachute near the town of Metz. Up until that point in the war, the only country to use parachute forces was France. Earlier in 1918, France had inserted small teams of demolition experts to raid German communications sites. Pershing did not think the plan could be implemented but gave Mitchell the go-ahead to continue planning for the Metz operation.

Mitchell returned to his headquarters and ordered Pershing's air priorities to be put into action. Then he and his operations officer, Major Lewis H. Brereton, began planning the parachute operation. They both knew that it would take until the spring of 1919 for there to be sufficient bombers and parachutes available to begin training the soldiers of the 1st Infantry Division in the use of parachutes. They also knew that the plan would be further complicated by having to use a large number of airfields to accommodate all of the airplanes that the plan would require. Additionally, once in the air, the airplanes would have to assemble en route to their drop zones.

None of this was easy to do. It was, indeed, a very complicated plan. Mitchell and Brereton were confident that it could work. Unfortunately for them, the war ended less than a month after Mitchell proposed his plan to Pershing.

However, early in 1919, Mitchell, newly promoted to Brigadier General and chief of all aviation activities for the U.S. Army, ordered the creation of a special board. This board was located at McCook Field, Ohio. Its purpose was to test and eventually select a parachute for army pilots. The board was headed by Major E. L. Hoffman. Parachute jumpers and designers throughout the country were invited to appear before the board and provide information and demonstrations.

The result of the board was a contract awarded to Leslie L. Irvin to provide sufficient quantities of his parachute to the army for use by its aviators. On 28 April 1919, Irvin demonstrated his parachute to the board. His design was similar to Broadwick's except that the parachute had no static line. It was a free-fall system, and on that day Irvin became the first man to make a free-fall jump from an airplane.

Irvin's system consisted of a back pack that was sewn to a harness that the jumper wore on his back. The pilot would jump from his airplane and, while falling, would pull a cord from the harness, opening the back pack and releasing a small parachute. Once this parachute, called a "pilot chute," deployed, it pulled the main chute with it. The pilot chute was connected to the apex of the main chute by a cord. The canopy of the main chute was connected from its skirt to the jumper's harness by suspension lines. The main parachute itself was circular and conical in shape. Irvin's system was also much more compact than Broadwick's design, making it more practical for use by pilots and crew-members.

Despite the advances in parachute design and practicality, the U.S. Army paid little attention to its use by soldiers other than aviators. It remained a life-saving piece of equipment. In 1928 and 1929, General Mitchell's concept was the object of two demonstrations employing parachuting soldiers. The first occurred in 1928 at Brooks Field, Texas, when a small squad was dropped. The following year, on 18 September, six soldiers jumped from a Martin bomber onto Kelly Field, Texas, assembled on the ground, and had their weapons prepared for use in about three minutes from the time they exited the airplane. None of the U.S. military observers at the Kelly Field demonstration took the concept seriously. Other countries, though, saw the value of being able to project infantry, artillery, and even tank units by delivering them onto a battlefield by means of a parachute.

Observers from both Russia and Germany were very impressed. Their observations and conclusions were put into action by their respective countries. Parachute concepts were developed and units formed in both countries, albeit clandestinely in Germany. It was even envisioned and demonstrated that infantry units thus deployed could capture airfields to facilitate subsequent air landing of additional units and equipment. Italy later joined Russia and Germany in developing parachute forces. Some of the planning by these countries was conducted under the cover of sports parachuting. Many civilian clubs were formed in the years between the two world wars in those countries, especially Germany, which was forbidden by the Treaty of Versailles from having certain forms or numbers of military formations. The members of these clubs formed the cadres of subsequent military units in those countries.

During the 1930s, the U.S. Army took no action to advance the military use of the parachute by ground units in terms of organizing, training, or employing such units. On occasion an article discussing parachute forces appeared in military publications but hardly anyone took notice. In 1937, such an article, entitled "The Employment and Defense Against Parachute Troops," appeared in the quarterly journal of the Command and General Staff College.

In early 1939, General George C. Marshall, the Army Chief of Staff, had been receiving reports that foreign militaries were organizing and training parachute forces. Marshall was receptive to the idea for the U.S. Army. Consequently, on 1 May of that year, he sent a memorandum to Major General George A. Lynch, the Army's Chief of Infantry. The memorandum directed that Lynch conduct a study to determine "the desirability of organizing, training, and conducting tests of a small detachment" of what Marshall called "air infantry." Marshall believed that such forces could be used to seize airfields on which subsequent forces could be brought in "by transport airplane." The study became bogged down for over six months by other considerations that the infantry believed had a higher priority. The study finally recommended that a provisional force of planes, probably about nine, be made available for such a study. Marshall passed this recommendation on to his Chief of Air Corps, Major General Henry H. "Hap" Arnold. The Air Corps was also loaded down with many other considerations and essentially decided that it did not have sufficient aircraft available to conduct further studies or tests. Marshall went along with Arnold's recommendations.

In Germany, an American infantry officer watched the subsequent developments and use of parachute forces and became interested in the concept. When he was transferred back to the War Department Operations Division, Major William C. Lee began talking about his new interest until he was eventually told to stop, that he was nothing but a nuisance. Following Germany's invasion of Norway, Belgium, and The Netherlands in May 1940, in which parachute and glider forces were employed to great effect, Major Lee was requested to brief President Franklin D. Roosevelt at the White House. The president wanted to know what Lee had seen while he was stationed in Germany. Following the briefing, the president directed the army to begin planning and training airborne units.

The glider

Man has always been intrigued by flight. Early Greek myths talk about the failed flight by Icarus and Daedalus in their escape from a prison on the island of Crete. Another Greek myth tells of Bellerophon and his flying horse Pegasus in their fight with the Chimera. Inevitably, some sought to fly in machines. These were machines that were not powered; the concept of powered flight came later in the development of flying. The first designs were mostly machines that "blew" on the wind. Some sources attribute early studies of flight to China and the invention of kites sometime in the 5th century B.C. Other sources cite designs made in the 1480s by Leonardo da Vinci. However, neither of these design efforts produced a successful working glider.

French brothers Joseph and Jacques Montgolfier invented the first hot air balloon. As with Jean-Pierre Blanchard's experiment, the first passengers in the Mongolfiers' hot air balloon were animals—a sheep, a rooster, and a duck in an experiment conducted in 1783. The first men to ride in one of the Montgolfier balloons were Jean-Francois Pilatre de Rozier and Francois Laurent, who ascended later that same year, on 21 November.

One of the men who contributed very much to the study of flight and the development of gliders was an Englishman, Sir George Cayley. He worked on his designs, experiments, and writings between 1799 and 1850. He invented the first glider capable of carrying a human aloft, although not for very long. Over the years Cayley altered his design many times, concentrating on using air flow to provide lift to his machines and the ability of the flier to move his body to control the machine. He also determined that a machine was needed to power his glider in order to achieve sustained flight.

The man credited with designing and inventing a glider that was capable of carrying a human for long distances was Otto Lilienthal, a German. In the 1850s, he and his brother Gustav began their experiments in the study of "the buoyancy and resistance of air." These experiments succeeded in 1891 when Lilienthal built and flew his first glider that was capable of carrying a man aloft. Lilienthal's model was able to achieve flight by having the man it carried take off by running downhill. In 1889 he wrote a book dealing with aerodynamics. This book and one written several years later by the French-born engineer Octave Chanute were used by Wilbur and Orville Wright to construct their early gliders.

In 1891, Samuel Langley, an American astrophysicist and astronomer, flew a model that he had designed that relied on an engine for power. It flew for three-quarters of a mile, when it ran out of fuel. Langley later designed a full-size version of his model but it turned out to be too heavy to fly. Disappointed, Langley turned to other pursuits and did not study flying any longer. He is, however, credited with the idea of "adding a power plant to a glider."

Elsewhere in America, by 1896 Octave Chanute built gliders based on a design of his own. His design was more stable than the Lilienthal model. So much more stable, in fact, that his gliders were capable of making 2,000 or more flights without an accident. In 1902, Wilbur and

Orville Wright built a successful glider that differed drastically from the models of Lilienthal and Chanute.

Previous glider models relied on, among other thing, a horizontal tail assembly. The Wright brothers' design was based on a vertical rudder that could be adjusted in flight. Their design also incorporated wings with flaps, capable of up-and-down movement. This led to a more stable craft, which was then used as the basis for their later powered flight. In 1911, following their pioneering work on flying powered aircraft, the Wright brothers returned to work on gliders.

Following World War I, the Treaty of Versailles banned Germany from using powered aircraft. Gliding became more prominent in Germany and many other countries. Sports clubs began to appear and national and international competition in gliding began to take place. At first, gliders were launched from high points or by running downhill and then soaring on wind updrafts as a means of remaining aloft. Newer methods for launching began to appear in the 1920s, such as rubber band launching and winch launching.

Rubber band launching essentially relied on aligning the glider's front wheel in a trough, with the very long rubber band routed through the hook at the front of the craft. The tail of the glider was either held back by ground crew or tied down. The band ran out to the sides of the glider, with a group at each end pulling the band out and to the rear. Once a certain tension was reached, the tail was released or the tie-down was cut. The front wheel popped out of the trough, the side ends of the band were released, and the craft was able to achieve sufficient energy to fly. Winch launching involved using a winch that was firmly set on a heavy vehicle. The glider was connected to the vehicle by a long steel-wire cable. The vehicle drove until the glider achieved a specific height and then the cable was released from the glider.

In 1925, the Treaty ban was partly dropped and Germany began experimenting with towing and launching gliders from airplanes. This method involved being able to tow the gliders for some distance before being launched by releasing the tow cable. There were even some experiments involving towing gliders by automobiles to launch them into the air, similar to the winch launching system but using a lighter vehicle.

Again, in Germany and Russia, formation of military units under the guise of civilian gliding clubs took place as a means of initiating research and testing the military applications of using glider aircraft as a means of delivering units and equipment to battle zones. The U.S. lagged behind in this research, relying initially on the tactic of airlanding transport (and sometimes bomber) aircraft to deliver troops and equipment, and then as a means of resupplying those units already on the ground in areas with secured airfields.

The first use of gliders in combat was on 10 May 1941, when German *fallschirmjägers* captured Fort Eban Emael in Belgium, previously thought to be impregnable. One consequence of this was that several other countries, including the U.S., began to reconsider the use of gliders for combat.

Sources

An earlier version of this chapter appeared on Military History Online: https://www.militaryhistoryonline.com/WWII/AirborneUnitsInWWII.

Books

Adleman, Robert H., and George Walton. *The Devil's Brigade*. Philadelphia, PA: Chilton Books, 1966.

Bergen, Howard R. *History of 99th Infantry Battalion, U.S. Army*. Oslo, Norway: Emil Moestue A-S, 1945.

Biggs, Bradley. *The Triple Nickels: America's First All-black Paratroop Unit*. Hamden, CT: Archon Books, 1986.

Blair, Jr., Clay. *Ridgway's Paratroopers: The American Airborne in WW II*. New York: Doubleday, 1985.

Brereton, Lewis H. *The Brereton Diaries: The War in the Air in the Pacific, Middle East and Europe, 3 October 1941–8 May 1945*. New York: William Morrow and Company, 1946.

Breuer, William B. *Geronimo! American Paratroopers in WWII*. New York: St. Martin Press, 1989.

Burhans, Robert D. *The First Special Service Force: A War History of the North Americans 1942–1944*. Washington, D.C.: Infantry Journal Press, 1947.

De Trez, Michel. *First Airborne Task Force: Pictorial History of the Allied Paratroopers in the Invasion of Southern France*. Belgium: D-Day Publishing, 1998.

Devlin, Gerald M., and William P. Yarborough. *Paratrooper! The Saga of the U.S. Army and Marine Parachute and Glider Combat Troops During World War II*. New York: St. Martin's Griffin, 1986.

Flanagan, Jr., Edward M. *The Angels: A History of the 11th Airborne Division 1943–1946*. Washington, D.C.: Infantry Journal Press, 1948.

Guthrie, Bennett M. *Three Winds of Death: The Saga of the 503rd Parachute Regimental Combat Team in the South Pacific*. Chicago: Adams Press, 1985.

Huston, James A. *Out of the Blue: U.S. Army Airborne Operations in World War II*. West Lafayette, IN: Purdue University Studies, 1972.

Raff, Edson D. *We Jumped to Fight*. New York: Eagle Books, 1944.

Rottman, Gordon L. *US Airborne Units in the Pacific Theater 1942–45*. London: Osprey Publishing, 2007.

Updegraph, Jr., Charles L. *U.S. Marine Corps Special Units of World War II*. Washington, D.C.: History and Museums Division, HQ US Marine Corps, 1972.

CHAPTER 4

The Force at la Difensa

Italy, early December 1943. It had been raining since mid-September. Rivers in the area were running high, bridges were swept away, and road surfaces were mostly gone. And, of course, it was cold.

The German Winter Line had held out despite attack after attack by Lieutenant General Mark Clark's Fifth Army. Regardless of any progress made, no advance beyond the Mignano Gap to Cassino was achieved. This Gap was flanked by the Camino hill mass including mountains such as la Difensa, la Remetanea, Rotondo, and Lungo.

On 22 November Fifth Army had announced Operation *Raincoat*, "the plan to breach the mountain passes." One of the units in this plan was new to the Italian theater, having previously served in the invasion of Kiska in the Aleutian Islands of Alaska. It was assigned to the U.S. 36th Infantry Division as the spearhead of the operation. This unit was the First Special Service Force.

The Force was created as a result of an idea by a British scientist named Geoffrey Pyke. In the summer of 1940, Pyke advocated for a low-silhouetted, tracked, propeller-driven sled designated the Weasel, which would be used by airborne troops in a counterinvasion of Norway. The idea, nicknamed Project Plough, caught the attention of Winston Churchill and was sent to the U.S. Army for review. Ground tests conducted by the 87th Mountain Infantry Regiment at Fort Lewis, Washington, proved very negative, although much of this was caused by a lack of snow that essentially resulted in the tests being inconclusive. Retests in the Sierra

Nevada range and on Mount Rainier again produced similar failed results particularly in the Weasel's capacity to transit a 20 percent grade.

In the War Department, the push was on to rush the project through in 180 days. Despite this, the study was thoroughly reviewed by Lieutenant Colonel Robert T. Frederick on the army staff. Frederick, a Coast Artillery officer who began his military career with the California Army National Guard, thought the project was a bad idea due to a lack of transport and an effective evacuation plan: "Plough was a beautiful paper concept but a strategical farce." He was, however, overruled, and the army, in its infinite wisdom, appointed Frederick to command a joint U.S.-Canadian unit that would train on and use the Weasel because he knew more about the project than anyone else. Frederick, by now promoted to colonel, was eventually sent to Fort William Henry Harrison, in Helena, Montana, to raise and train the unit.

Frederick decided that his unit would have to be made up of out-of-the-ordinary soldiers and began looking for officers to fill the staff of his organization. A request went out to all army units seeking volunteers with specific backgrounds or skills. It read, in part, that Frederick was looking for:

Any soldier, officer or enlisted, selected would be advised that he would be required to complete parachute training.

As recruiting continued in the U.S. Army, meetings were held with Canadian Army representatives to settle problems of organization, pay, uniforms, records administration and maintenance, quarters, and discipline. In a short time, arrangements were made to recruit soldiers from the 1st Canadian Parachute Battalion; those chosen were redesignated as the 2nd Canadian Parachute Battalion, which existed only within the Force.

In July 1942, soldiers began to arrive at Fort Harrison and, within 48 hours, began their parachute training. Frederick's training concept for his unit required all soldiers to "master all the infantry small arms, be able to drive and repair the Weasel, be a qualified skier and parachutist as well as being a master of demolitions and numerous lesser skills." Training began in earnest, starting with small unit tactics, with plans to move on to integrated unit tactics, and to such specialized training as "skiing, rock climbing, living in cold climates …" as well as operations associated with driving and maintaining the unit's snow vehicles. In late August, Frederick awarded 1,200 parachute qualification badges to the combat echelon of his unit. The Force consisted of three regiments with two battalions each; each battalion was divided into three companies. The Force also included a service battalion, composed of clerks, cooks, mechanics, armorers, riggers, supply specialists, and medics who would perform all work details, leaving the combat echelon to train continuously. Canadian and U.S. soldiers, both officer and enlisted, were equally integrated throughout the Force.

Naming his organization presented an interesting thought process for Frederick. Various designations were suggested that incorporated such words as "Parachute Infantry" or "Commando" or "Ranger" or other "tough-sounding" names. The term "Plough" was considered to be a secret operational name for the unit and so would not be used. Frederick wanted something less sinister and he eventually decided on the innocent sounding "First Special Service Force." The Quartermaster Corps' Heraldry Section assisted in the selection of a shoulder patch for the Force by designing a red arrowhead with "USA-Canada" across the top and down the center. Since his soldiers, both U.S. and Canadian, were selected from all branches of their respective militaries, Frederick wanted a separate insignia for their uniforms. Eventually, with the formal written agreement of the remaining members of the U.S. Army's Indian Scouts, the collar insignia of crossed arrows was selected. The War Department formalized this selection, thereby creating the Force as a separate branch within the army.

While the administrative details were being worked out, Frederick looked to organizing and training his Force. Lieutenant Colonel John G. McQueen, of the Calgary Highlanders, had led the Canadian element to Fort Harrison. He was designated the Executive Officer of the Force. Two of the regiments were commanded by Americans with Canadian executive officers; a Canadian officer, with an American executive officer, commanded the other regiment. Of the six battalions in the Force, five of them were commanded by Canadian officers with American executive officers. More Canadian officers than American officers commanded

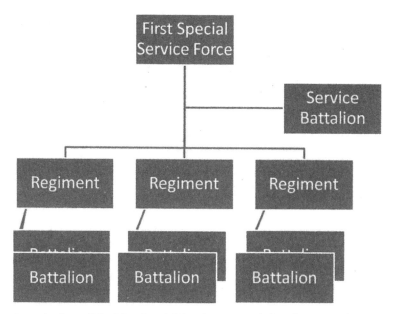

Organization of the First Special Service Force. (Chart by Dave Shroyer)

companies; however, most platoon commanders were American first sergeants. Within the ranks of the companies, the mix of nationalities was essentially based on the proportion of the two nationalities within the total Force structure.

Training began according to a tight schedule. This schedule consisted of three phases—3 August to 3 October 1942 for parachute training, basic weapons qualifications, demolitions training and qualification, small unit tactics, and a lot of physical training; 5 October to 21 November 1942 for more tactics and field problems; and 23 November to late December 1942 for ski training, rock climbing training, cold climate survival training, and operations with the combat snow vehicle. Training days ran from 0445 until 2130 four days of the week, with the day's training shortened until 1700 on the remaining two days. The men of the Force did not train on Sundays.

Planning was conducted at the same time as training. The Force S-2 (Intelligence) and S-3 (Operations and Training) staffs were reviewing plans submitted by Washington when the Force was committed to combat.

Missions included attacking and destroying German oil fields near Ploesti, in Romania, as well as power generation capabilities in Italy. When these planned operations were reassigned to air assets, planning returned to targets in Norway and elsewhere in the Scandinavian countries.

Late in the summer, Lieutenant Colonel McQueen broke his leg during a training parachute jump and was transferred out of the Force. He was replaced by Colonel Paul D. Adams, an American. Meanwhile, hand-to-hand fighting was added to the already jammed training schedule. At the suggestion of former Shanghai Police Captain Bruce Fairbairn, the Force obtained the training services of another former member of the Shanghai Police Department, an Irishman named D. M. O'Neill. The so-called "O'Neill System" of hand-to-hand combat was based on what O'Neill called "a kick and a poke," a simplified yet more destructive system over Judo.

While Frederick was on a coordination visit to London in September 1942 to secure sufficient basing areas for his Force in England, as well as air assets to move it on deployment, he was notified to suspend his unit's current planning. Project Plough had been cancelled.

Training continued despite the demise of the Plough plan, and now the Force was directed to make the unit "into a versatile assault group able to undertake any task that might be assigned." The Force acquired Johnson light machine guns from the U.S. Marine Corps by trading a part of its store of a new explosive known as RS, developed by the Army Ordnance Department. The newest in cold winter equipment and clothing came from the Arctic Experimental Department of the Quartermaster Corps. Light weight, long range radios were tested.

In late November, Frederick was alerted to cancel plans that he had previously been given for the Force to be used in the Caucasus Mountains and in New Guinea. This time, the target for planning was Kiska, in the Aleutian Islands. This island chain was partially occupied by the Japanese. The target date for the invasion was the spring of 1943. In January 1943, Frederick visited Alaska and his Force was accepted into the Arctic Experimental Department for the invasion. In late March, the Force was scheduled to receive amphibious training the following month in Norfolk, Virginia. The unit left Fort Harrison and headed east.

More training followed this, all to make the Force more versatile and ready for deployment. When their training at the Navy Scouts and Raiders School in Fort Pierce, Florida, was cancelled, the Forcemen were sent to Fort Ethan Allen, Vermont, for more cold weather exercises. The Force left Fort Ethan Allen by train on 26 June and arrived in the San Francisco area by 3 July.

On 9 July, the Force loaded onto two Liberty ships bound for Adak, an island in the Aleutians, where the men were met by Frederick. He had spent the previous week becoming familiar with his unit's bivouac and pre-invasion training areas. He had even tagged along on a bombing mission over Kiska to examine the Force's target areas. In the meantime, men of the Force were meeting with members of the Alaska Scouts, who were able to give them local information about their objectives.

The invasion plan called for the 1st Regiment to be part of the initial landing force on the southern end of the island and then fight its way eastward toward the Japanese naval base at Kiska Harbor. The following day, the 3rd Regiment was to land on the western side of the island and to fight eastward, forming a pincer against the Japanese forces. Frederick and his staff were to accompany the 3rd Regiment. The 2nd Regiment was to be held in reserve, to be parachuted in where tactically needed. The invasion was to kick off on 15 August.

As the first wave went in at 0120 on the 15th, they found, to their utter surprise, that the Japanese had left the island just a few hours before. The landing units proceeded eastward, still finding no Japanese. By 0300 the following morning the landing force on the western side was ashore, and still did not find any Japanese. As a result, the parachute jump was cancelled and, later that day, the island declared secured. Two days later, on the 18th, Frederick was directed by a wire from Pacific Fleet Headquarters at Pearl Harbor to return the Force to San Francisco "without delay." The Force returned on two ships, one arriving on 30 August and the other on 1 September.

Half of the Force was given 10 days of leave, with orders to report back to Fort Ethan Allen. The other half left San Francisco by troop trains, bound for Vermont. Once everyone returned from leave, the Force spent a month tightening up and preparing for deployment overseas. On 14 October, orders came for the Force to move by train to Camp Patrick Henry, Virginia, and then to the Hampton Roads Port of Embarkation for overseas deployment by ship. The trip would take several weeks. Their destination was Casablanca, then on to Oran and Naples. In just three months, the Force had gone from the Aleutians Islands of Alaska to face the Japanese to the mountains of Italy to face the Germans. General Dwight Eisenhower began developing plans for the employment of the Force.

In sending for the Force, it had been General Dwight Eisenhower's intent to use them for raids, sabotage, or guerrilla operations in Italy, or perhaps in southern France or the Balkans. But General Mark Clark, Commanding General, Fifth Army, beat him to the punch and had the Force assigned to his command.

Typically, November is the rainiest month in Italy. November 1943 was no exception. In fact, it had been raining fairly steadily since September and was getting colder as winter approached. The so-called German "Winter Line" was actually a chain of heavily fortified hills manned by some of the best units the Germans had in Italy, including the Panzer Grenadier Division and the Herman Goering Division. These units had held out on Monte la Difensa and Monte la Remetanea against everything the Fifth Army had sent against them. The Allies suffered heavy casualties in each attack, particularly in a sustained drive in the 12 days prior to the Force being assigned to the area.

One of the keys to the Allies getting through the Winter Line was the 3,000-foot-high Monte la Difensa. Air and artillery attacks had not been successful. Far above the 1,000-foot tree line, the Germans were dug in among sheer cliffs. Connected, concealed trails along the upper part of la Difensa made it easy for the Germans to maneuver and coordinate their defense. The cliff face actually begins at about 2,000 feet and extends upwards at a pitch of between 60 and 70 degrees for almost another 1,000 feet.

The Force was designated the next spearhead against la Difensa and la Remetanea. The planners were not optimistic that the Force would be able to succeed where other units had failed and, even if they did succeed, would probably not be a combat effective unit afterwards.

During the first week after being given the assignment, Frederick and a small group of enlisted men and a few members of his staff conducted night-time reconnaissance of the targets. The northern approach consisted of a 200-foot cliff. Above this were a series of six ledges about 30 feet apart. Frederick found that the Germans apparently believed that a full-sized force would not be able to attack up this flank. As a result, there was very little German firepower employed to protect against an assault from this direction. Frederick decided that this was the approach the Force would use.

By 15 November, the Force was in position and ready to launch. The rain continued to come down heavily and it was cold. The Force planners believed that these conditions would add to the surprise of their attack. The 2nd Regiment was designated as the lead attacking element to take the objective. The 1st Regiment was to serve as the reserve unit for the 36th Division. The 3rd Regiment was designated to serve as supply carriers, litter bearers, and Force reserve. D-Day was set for 2 December 1943.

At 1600 on 2 December, the men of the Force began loading trucks and moved forward. In heavy rain, they left the trucks and hiked the last 10 miles to the base of their objective. The 2nd Regiment began climbing under the cover of rain and darkness. Their objective that night was a point about halfway up to the crest. They were not in a hurry and stealth was their method of movement that night. Artillery fire covered their movement and, they hoped, kept the Germans busy protecting themselves and not concerned about their unprotected flank. As the lead element moved up the mountain, they left ropes in place for the double row of companies coming behind to use. These ropes were also used to evacuate the wounded and other casualties down the mountain later in the action. Before first light all elements of the Force were in their designated positions. They "burrowed into concealment" and remained there all day. They had not been detected by the Germans.

During the day down below, other Allied units moved up to their assigned positions. This was a contingency movement. If the Force was successful in securing its objective, the other units would be needed to reinforce the victory and prevent German reinforcement or counterattack.

Late that afternoon, the men of the 2nd Regiment made a final check of their weapons and equipment. Frederick led the Force elements as they began to climb the trail upward. The goal for the 2nd Regiment that night was to arrive at the base of the crown of la Difensa by 2230.

The Forcemen began their rigorous climb, carrying weapons and heavy backpacks. They crawled upwards from hand-hold to hand-hold in the cold rain, using ropes whenever possible to facilitate the climb. By midnight most of them were well concealed in the rocky ledges below the German positions. There was no indication that the Germans knew of their presence or even of their approach. They tied their ropes around themselves and rested in the darkness, the cold air freezing their sweat.

Soon Frederick and some of his staff joined the weary Forcemen in the lead. The ledge where they waited was barely the width of a man's body. They could smell food odors coming from the German mess area; they were that close to the enemy yet still undetected. Sometime around 0430, Frederick gave the signal for his men to move forward. Bayonets and knives were the order for this approach. Frederick later recalled that "Fog shrouded the movement forward of the scouts." The only sound the Forcemen made as they moved into the German position was a "soft gurgle vented by German sentries who had their throats cut" by the leading Forcemen.

Suddenly several of the Forcemen slipped on loose stones that the Germans used around their enforcements. A green flare went up, then a red one. Then two bright magnesium flares lit the sky, catching the advancing Forcemen in their light. The Force had been spotted and the battle was on.

The Forcemen were attacking from an advantage, even though silhouetted in the flares. They had come from a position which the Germans were not expecting. The Germans recovered in one area and brought extra machine guns on the advancing Americans, slowing them down. After being pinned down for a short time, a nine-man group, with machine gun and mortar coverage, advanced and captured a cave and dispatched a nearby machine gun.

As light began to spread, Germans began to surrender. At one point, a German soldier advanced while waiving a white flag. A Force company commander moved out to take the prisoner. As soon as the captain was exposed, the German dropped into a hole and other Germans nearby opened up on the captain. The outraged Force company ran forward and killed all the Germans. Thereafter that company made no attempt to capture prisoners.

Two companies were now on top of la Difensa. From this position they could see the Germans "swarming like ants toward the valley and next hill [la Remetanea]." By 0700, 1st Battalion occupied the hill top. The men of 2nd Battalion arrived soon after. It took about another hour to clear the area of Germans and dig in, waiting for the expected counterattack.

In the meantime, the 3rd Regiment and part of the service battalion combed the area for wounded and began moving them down the mountain as quickly as they could. One of the battalion commanders noted that the final assault had been up an almost perpendicular wall, one, however, with several ledges.

A request went down the mountain for ammunition as those Forcemen on top were low. Fortunately, no counterattack was launched right away. The Germans were satisfied, for the time being, to rain mortar shells on the enemy atop the mountain. Frederick, after reviewing the situation, sent a message down requesting urgent resupply of ammunition, rations, water, blankets, and litters. It took about six hours for this material to move from the bottom of the mountain to the top.

In other areas, British units were advancing on their objectives on nearby hill masses. Later in the day, with no German counterattack still in sight, the 36th Division inquired about Frederick's intentions as some adjoining units were taking heavy fire from some of the nearby hills. In the afternoon, Frederick approved plans for the 2nd Regiment to attack la Remetanea the following morning, on the condition that patrols continue during the night and an artillery forward observer be kept on alert and well forward to call in fire support as soon as he deemed it was necessary.

Information brought back by various patrols caused the attack to be delayed for two more days, allowing for more consolidation of friendly forces so as to overwhelm the Germans. On 7 December 1943, the

Germans were driven from la Remetanea, with the Force in the forefront of the action. The fate of the German defense in the immediate hill area was sealed. The Winter Line had been broken. The First Special Service Force had earned its first battle success in an attack that took the German defenders completely by surprise.

Later, General Eisenhower described the battle like this: "... a small detachment [the Force] had put on a remarkable exhibition of mountain climbing. With the aid of ropes, a few of them climbed steep cliffs of great height. I have never understood how, encumbered by their equipment, they were able to do it ... They entered and seized the [German company commander], who ejaculated, 'You can't be here. It is impossible to come up those rocks.'"

The Force would move on to other victories. They would take part in the landings at Anzio (where they would earn their nickname of "Black Devils" from the German units opposing them), be among the first units into Rome, and land in the invasion of Southern France in August 1944. They would eventually be disbanded in December 1944 with the Canadians returning to Canadian units and the Americans to American units, having earned many battle credits. They succeeded where many others had failed and satisfied all of Colonel Fredericks's expectations of them.

Sources

Books

Adleman, Robert H., and George Walton. *The Devil's Brigade*. Philadelphia, PA: Chilton Books, 1966.

Burhans, Robert D. *The First Special Service Force: A War History of the North Americans 1942–1944*. Washington, D.C.: Infantry Journal Press, 1947.

De Trez, Michel. *First Airborne Task Force: Pictorial History of the Allied Paratroopers in the Invasion of Southern France*. Belgium: D-Day Publishing, 1998.

Eisenhower, Dwight D. *Crusade in Europe*. New York: Doubleday & Company, Inc., 1948.

Hicks, Anne. *The Last Fighting General: The Biography of Robert Tyron Frederick*. Atglen, PA: Schiffer Military History, 2006.

CHAPTER 5

Sabotaging Hitler's Heavy Water

The years leading up to and including World War II saw a race by Germany and the United States to develop an atomic weapon. Although the idea of nuclear fission was first mentioned in 1934, it was not until four years later that experiments confirmed it by using Uranium. The two methods for moderating the energy of neutrons loosed by bombarding Uranium involve the use of heavy water or graphite. Heavy water, or Deuterium, which looks like regular water, was discovered in 1933. Germany ultimately decided to use heavy water in its nuclear reactor to breed the Plutonium-239 needed in its weapons research.

One method of producing heavy water is by separating it from regular water using electrolysis. This method requires electrolysis chambers and a considerably large amount of power. Ultimately the heavy water supplier for scientists throughout the world was the hydroelectric plant run by Norsk Hydro, located near Rjukan in the Telemark region of Norway.

By late January 1940, Germany had begun procuring heavy water from Norsk Hydro through the firm of *IG Farbenindustrie Aktiengesellschaft*. At about the same time the French Military Intelligence organization, *Deuxieme Bureau*, conducted an operation to remove what was then the world's supply of heavy water from Norsk Hydro. This amounted to about 44.25 gallons. In a risky operation the heavy water was removed from Norway and sent, ultimately, to Falmouth in England by a roundabout route (Oslo to Perth in Scotland, to Paris, to Bordeaux, and to Falmouth).

Despite this removal, Allied commands and intelligence organizations believed that the Germans would use Norsk Hydro to continue its heavy water production operations to produce a sufficient supply for use in their weapons research program. The British Special Operations Executive (SOE), charged with conducting, among other operations, sabotage campaigns, began planning several operations aimed at limiting or destroying heavy water production at Norsk Hydro.

The first phase, codenamed Operation *Grouse*, was conducted on the night of 18–19 October 1942. In this operation, four members of the Norwegian resistance (Knut Haugland, Arne Kjelstrup, Jens-Anton Poulsson, and Claus Helberg) parachuted onto the Hardanger plateau, which was above and near the Norsk Hydro plant. Their mission was to gather as much intelligence (including blueprints, if possible) concerning the plant and its heavy water operations.

The information that the *Grouse* team obtained was sent back to SOE for use in planning the next phase of operations. This phase was codenamed Operation *Freshman*. For this operation, a force of sappers (combat engineers), all paratroopers of the British 1st Airborne Division, was sent in by gliders to link up with the *Grouse* team. The combined teams was to set and detonate demolitions in key parts of the Norsk Hydro plant to destroy its heavy water capability, then escape and evade through Norway to Sweden. The *Freshman* team left England on the night of 19 November 1942.

Weather conditions in the vicinity of the designated drop zones on that night were extremely poor. As a consequence, the link up of the two teams did not take place. One of the Halifax bombers towing a glider crashed into a mountain, killing its entire crew. Shortly before, the glider attached to this tug was able to detach itself, but crash landed, resulting in many casualties among the sappers. The other glider also crash landed, killing most of its occupants. The second Halifax limped back to England and landed safely. All of the surviving sappers were captured by the Germans, tortured, interrogated, and eventually murdered. This was in keeping with what was called the Commando Order, issued by Adolf Hitler on 19 October 1942. The Commando Order stated that all Allied commandos that German forces captured would be killed immediately, regardless of whether they were in uniform or tried to surrender. Failure to carry out this order was considered an act of negligence by German military law.

Because of the *Freshman* operation, the Germans became aware of Allied interest in the heavy water operations at Norsk Hydro. Plans were immediately put into effect to upgrade the perimeter defenses at the plant. This included establishing a mine field around the plant, putting in floodlights on the perimeter, and increasing the size of the guard force. The increased security did not deter SOE.

Planners searched for another way to send in commandos to link up with the *Grouse* team, which by now had been renamed the *Swallow* team. The four commandos of *Swallow* team lay low for several months in the mountains above the hydro plant, continuing to maintain contact with SOE in London by radio. They subsisted as best they could on whatever was available. One report states that they were able to kill a reindeer in late December 1942.

SOE finalized the plans for the next phase, codenamed Operation *Gunnerside*. In this phase, six additional Norwegian commandos (Joachim Ronneberg, Knut Haukelid, Fredrik Kayser, Kasper Idland, Hans Storhaug, and Birger Stromsheim) parachuted into Norway on the night of 16–17 February 1943. Several days later, the *Gunnerside* team linked up with the *Swallow* team and they began planning for their assault on the plant. They decided to attack on the night of 27–28 February 1943.

In the intervening months after the failed Operation *Freshman*, security had become somewhat slacker. The lights and mine field remained in place, but the guards had become complacent. Nevertheless, the only bridge spanning the 246-foot-wide ravine over the Maan River was still guarded at full strength. Because of this, the combined teams decided that their best approach was to climb down the 656-foot-deep ravine, cross the icy river, and then climb up the other side of the steep ravine. Using the information obtained about the plant layout, the teams found and followed a single rail line into the plant area without being detected by any of the guard force.

Once inside the plant itself, the teams encountered a Norwegian caretaker who assisted them in their subsequent movement in the plant.

The teams moved quickly to the area where the heavy water electrolysis chambers were located and put their explosives in place. A long time-delay fuse was attached to the explosives and the teams prepared to light it and leave. They left behind a British submachine gun, hoping the Germans would draw the conclusion that the sabotage was the work of British forces. They hoped that this would prevent any German reprisals against the local citizens. Just as they were ready to light the fuse, the caretaker stopped them; he could not find his eyeglasses! Since eyeglasses were very difficult to get, the commandoes conducted a search of the area and found them. They then lit the fuse and left the plant the same way they had entered.

After they were clear of the plant, the explosives detonated. The heavy water electrolysis chambers were destroyed as was almost 120 gallons of heavy water stored on site. The Germans launched an extensive search operation for the commandos but did not find any of them. Some escaped to Sweden, some to Oslo, and a few remained in the general region. Knut Haugland, of *Swallow* team, was eventually the only member of the two teams who remained in the immediate vicinity.

Although the operation was deemed to be a success, the Germans were able, over several months, to repair some of the damage and restart production of heavy water in April 1943. SOE considered conducting another raid on the plant but dropped the idea as too difficult. In November 1943, the US Army Air Forces were ordered to conduct a series of bombing raids on the plant. Although only about one in seven of the bombs dropped actually hit the plant, these raids resulted in extensive damage to the plant. Eventually the Germans decided to halt heavy water operations at the Norsk Hydro plant and move the heavy water on hand to Germany. Planning and execution of the move was conducted with very tight security. One phase of the movement required railway cars carrying the heavy water to transit Lake Tinnsjo by a ferry, the *SF Hydro*.

Knut Haugland learned of the plan to move the heavy water and began planning another sabotage operation. He eventually decided that the phase involving the ferry was probably his best opportunity. With the assistance of several members of the local resistance and a crew member on the ferry, Haugland was able to sneak aboard the ferry prior to its departure and place plastic explosives next to the keel, to be detonated by two alarm clock fuses. On 20 February 1944, soon after leaving its dock, the *SF Hydro* sank into one of the deepest parts of the lake, after the explosives ripped holes in its keel.

The missions of the various teams sent into Norway were ultimately successful. The German heavy water development program was finally stopped.

Knut Haugland eventually escaped back to England, after several narrow escapes from the Gestapo. For his actions in the raids, he was awarded the War Cross with Sword, Norway's highest decoration for gallantry. In 1947, along with another Norwegian (Thor Heyerdahl), who he had met at a training camp in England in 1944, and several others, Haugland took part in the Kon Tiki expedition, crossing part of the Pacific Ocean on a balsa wood raft from east to west. When Haugland died in December 2009, he was the last living member of the Kon Tiki expedition.

Sources

Books

Gallagher, Thomas. *Assault in Norway: Sabotaging the Nazi Nuclear Program*. Guilford, CT: Lyons Press, 1975.

Haukelid, Hans. *Skis Against the Atom*. London: William Kimber, 1954.

Mears, Ray. *The Real Heroes of Telemark: The True Story of the Secret Mission to Stop Hitler's Atomic Bomb*. London: Hodder & Stoughton, 2004.

Wiggan, Richard. *Operation Freshman: The Rjukan Heavy Water Raid 1942*. London: William Kimber, 1986.

Articles

Grimes, William. "Knut Haugland, Sailor on Kon-Tiki, Dies at 92." *New York Times*, January 3, 2010.

CHAPTER 6

The Alamo Scouts—LRRPs of World War II

In late November 1943, Lieutenant General Walter Krueger, Commander of U.S. Sixth Army, created a special reconnaissance unit to be at his disposal for scouting and raiding missions. The impetus for this move lies in two separate actions. First was General Douglas MacArthur's refusal to allow the Office of Strategic Services (OSS) to operate in the Southwest Pacific area but instead organized an agency, the Allied Intelligence Bureau, to conduct the same missions as the OSS. Second was the disastrous invasion of Kiska Island in the Aleutians. Intelligence in the latter operation was so bad that the invading force did not know that the Japanese had departed the target islands. Krueger did not want this to happen to him.

Krueger's order established the Alamo Scouts Training Center (ASTC) under the direction of the Sixth Army G-2, Colonel Horton V. White. At the time, Alamo Force was the code-name for Sixth Army. The order called for a six-week training course, to begin before the end of the year, and for all divisions in Sixth Army to supply soldiers to attend the training. Krueger picked Lieutenant Colonel Frederick W. Bradshaw to be the director of the ASTC. From among the graduates of the training course, Bradshaw was to select teams that would conduct the missions; those graduates not selected for the teams would return to their parent units to pass on the benefits of their training and to conduct similar missions.

During the course of the next two years, as the Sixth Army Headquarters moved forward so did the ASTC. In all, nine classes were conducted at five different locations. These nine classes included 75 officers and 250 enlisted men among their graduates. Not all of these graduates served in active Alamo Scouts teams, however. Some of the graduates remained as instructors or staff members at ASTC. The various teams ultimately comprised 21 officers and 117 enlisted men.

Training was divided into two phases. Phase one lasted four weeks and consisted of class-room and hands-on refresher training in a variety of subjects. "Refresher" is the proper term because this training was never intended to be any kind of basic course, rather it was to rely on all of the training that the soldiers had received to that time and to concentrate on those subjects that would be needed for their duties on Alamo Scouts teams. Throughout the training cycle each student had a chance to work with and evaluate all the others.

Phase two lasted two weeks and consisted of the practical application of the previous training to strenuous field problems, each lasting three to four days. These patrols required the teams to plan, brief, and execute the mission assigned. Support included U.S. Navy PT boats for approaches and small rubber boats for landings. Another team assisted with the infiltration and exfiltration phases of the patrol; this team was called the "contact team."

ALAMO SCOUTS TRAINING CENTER LOCATIONS

1. Kalo Kalo, Fergusson Island, New Guinea
    Director: LTC Frederick W. Bradshaw
        Class 1: 27 December 1943–5 February 1944
        Class 2: 21 February 1944–31 March 1944
2. Mangee Point, Finschafen, New Guinea
    Director: LTC Frederick W. Bradshaw
        Class 3: 15 May 1944–22 June 1944
3. Cape Kassoe, Hollandia, New Guinea
    Director: MAJ Homer A. Williams
        Class 4: 31 July 1944–9 September 1944
        Class 5: 18 September 1944–28 October 1944
4. Abuyog, Leyte, Philippine Islands
    Director: MAJ Homer A. Williams
        Class 6: 26 December 1944–1 February 1945
5. Mabayo, Bataan, Luzon, Philippine Islands
    Director: MAJ Gibson Niles
        Class 7: 23 April 1945–1 June 1945
        Class 8: 18 June 1945–6 August 1945
        Class 9: 6 August–2 September 1945
            (interrupted by the end of the war)

Classes and locations of the Alamo Scouts Training Center. (This is a variation of an illustration in *Silent Warriors* by Lance Q. Zedric. Variation and use with permission from the author)

As subsequent classes were conducted and the type of missions given to the operational teams changed over time, training was evaluated and based on how beneficial it was to the operational teams. There were minor changes in the curriculum, the biggest of which was to extend Phase two for an additional half week and cut some of the classroom training. Over the course of the two years that the ASTC existed, the training schedule changed very little.

The Alamo Scouts teams consisted of a lieutenant and five or six scouts. Generally, each team operated alone unless a mission called for additional scouts to go along or for more than one team to operate together or in the same nearby area. The teams were named after the officer who led them rather than by any numerical designation. The most interesting thing about the teams was how they were chosen. At the end of training, each enlisted man was asked to name three other men with whom he would go on a four-man patrol and explain why. The officers were asked to pick five men who they wanted on their team. By this time, everyone in the class had worked with everyone else and was fairly familiar with who was compatible with them. The ASTC faculty had the final say on who made an operational team, but their lists were very similar to those of the students. From the first day of training to the last, the concept of the team was drilled into the students.

Once they were operational, this concept became a reality for the Alamo Scouts. Every member of the team made a contribution to every mission. Just because the lieutenant was in charge did not mean he had the only ideas or the final say-so. Every man was expected to speak up when he had something to say.

The Alamo Scouts operated behind Japanese lines during the two years they existed. In many cases, behind the lines consisted of a series of islands, all held by the Japanese, which would be invaded by Sixth Army units in a series of amphibious attacks during 1944. Sixth Army G-2 needed to know not only about unit dispositions on these islands but also beach conditions, such as gradient angle, sand composition, water depth, obstacles, etc.

Most of the Alamo Scouts' missions were to conduct reconnaissance and gather intelligence, foreshadowing the LRRP patrols of the Vietnam War. Mission insertion (and later extraction) was by rubber boats launched

This photo was taken at Finschhafen, New Guinea, on 22 June 1944, the day that class three graduated from the Alamo Scouts Training Center there. Those pictured include staff and graduates. Front row, left to right: Lieutenants John M. Dove, Milton B. Beckwith, Donald P. Hart, Gean H. Reynolds, William B. Lutz, and Arpad Farkas. Back row, left to right: Lieutenant Lewis B. Hochstrasser, Captain Richard G. Canfield, Major Homer A. Williams, Captain Gibson Niles, Lieutenant Gus Watson (Australian Army), and Lieutenant Robert S. Sumner. (Lewis B. Hochstrasser; used with permission)

from Catalina (PBY) aircraft, submarine, or PT boat. Later, following the invasion of Luzon in the Philippines, Alamo Scouts teams worked closely with Philippine guerrilla units, much like some modern special forces operations. At least six of their missions were planned to capture Japanese prisoners, two were to liberate prisoners held by the Japanese, and several were to recover downed U.S. pilots, air crews, or sensitive equipment and code books.

Some examples of missions conducted by Alamo Scouts teams are as follows:

> McGowen Team: conducted in February 1944 on Los Negros in the Admiralty Islands. The mission was to gather intelligence for an operation to be conducted by elements of the 1st Cavalry Division which would conduct a reconnaissance in force prior to an amphibious landing.

Sumbar Team: conducted in May 1944 on Vandeomoear Island. The mission was to determine how many Japanese troops were located near Sarmi Harbor.

Thompson Team (in conjunction with Naval Amphibious Scouts and others): conducted in June 1944 on Wargeo Island. The mission was to locate three suitable sites on the island for the construction of air and naval bases.

Hobbs Team: conducted in July 1944 on Japen Island. The mission was to capture a merchant ship captain who had been supplying information about American forces to the Japanese; he was captured.

Sumner Team: conducted in August 1944 on Pegun Island. The mission was to reconnoiter the island and determine the possibility of recovering a PT boat crew that had been captured; there was no evidence that the crew were alive.

Nellist Team and Rounsaville Team: conducted in October 1944 at Cape Oransbari in New Guinea. The mission was to attack and recover 66 Prisoners of War (POWs) from a Japanese camp; the prisoners were all recovered alive.

Littlefield Team: conducted in December 1944 near Leyte. The mission was to gather intelligence on Japanese forces in the vicinity of the beach at Camo Downes, which was located west of Ormoc.

Nellist Team and Rounsaville Team (in conjunction with a company from the 6th Ranger Battalion and Philippine guerrillas): conducted in late January and early February 1945. The mission was to attack the Japanese POW camp at Cabanatuan on Luzon and liberate the 531 POWs held in custody; the prisoners were all recovered alive.

Chanley Team: conducted in February 1945 and later. The mission was to work with Philippine guerrilla teams in and around the Manila area.

When not on missions behind the lines, the Alamo Scouts pulled personal security duty for General Krueger as he visited units at or near the front. After the Japanese surrender, two teams of Alamo Scouts accompanied Sixth Army headquarters to Japan. There they conducted missions to identify and secure weapons caches and to continue personal security duty for General Krueger. In November 1945, the last of the Alamo Scouts teams were disbanded. During the more than 110 missions conducted by the Alamo Scouts, not one of the Scouts was killed or captured—an incredibly impressive achievement.

The Alamo Scouts served as the pattern for similar units in later wars. The Vietnam War's Long Range Reconnaissance Patrols (LRRPs), Long Range Patrols (LRPs), 75th Ranger companies, and later the

Sumner Team graduation from the Alamo Scouts Training Center, 22 June 1944. Front row, left to right: Private First Class (PFC) Paul B. Jones, PFC Edward Renhols, and Corporal William F. Blaise. Back row: Lieutenant Robert S. (Red) Sumner, Staff Sergeant Lawrence E. Coleman, PFC Harry D. Weiland, and Corporal Robert Schermerhorn. (Robert S. Sumner; used with permission)

Long Range Surveillance Units have many of their aspects drawn directly from the Alamo Scouts. The training schedule for the MACV Recondo School was very similar in composition to that used by the ASTC. As a result, special forces considers the Alamo Scouts as part of their heritage. In December 1995, when the U.S. Army Special Operations Command dedicated its new headquarters building at Fort Bragg, North Carolina, the adjoining U.S. Army Special Operations Forces (ARSOF) Memorial Plaza contained a memorial marker of the Alamo Scouts to reinforce this heritage. In addition, the U.S. Army Ranger School adopted the ASTC principle of sending graduates back to their parent units to pass on the benefits of their training to other members of their unit.

Some of the members of the Rounsaville and Nellist teams on the day after the 30 January 1945 raid on the POW camp at Cabanatuan. Those pictured are, left to right, front row: PFC Galen C. Kittleson, PFC Rufo Vaquilar, Lieutenants William E. Nellist and Tom J. Rounsaville, and PFC Franklin Fox. Back row: PFC Gilbert Cox, Technical Sergeant Wilbert Wismer, Sergeant Harold Hard, Corporal Andrew Smith, and PFC Francis Laquier. Missing from the photo are Staff Sergeant Tom Siason, PFC Sabis Asis, and Technical Sergeant Alfred Alfonzo. (Photo courtesy of the Alamo Scouts Association; used with permission)

Of all the stories from all the missions, probably the one which best demonstrates the close relationship between the Alamo Scouts and General Krueger, their founder, is from a mission that was cancelled. In June 1944, Lieutenant Woodrow H. Hobbs and his team were told to plan the kidnapping of two Japanese generals. A Japanese prisoner, who had worked at the camp where the generals lived, had provided very detailed information about the camp layout and the daily routine of the generals, such as when they ate, when one of them went horseback riding, and so forth. The Hobbs Team, working with training center operations officer Lieutenant John R. C. McGowen, briefed Krueger on their plan. The plan was feasible, but the mission would have been extremely dangerous. There were limited ways in, and these were always under observation. The generals were well guarded, and there was every likelihood that the team would take serious casualties. Krueger cancelled the operation and made the terse statement: "I wouldn't take the whole damn Jap army for one Alamo Scout."

Sources

Books

Breuer, William B. *The Great Raid on Cabanatuan: Rescuing the Doomed Ghosts of Bataan and Corregidor*. New York: John Wiley & Son, 1994.

Hockstrasser, Lewis B. *They Were First: The True Story of the Alamo Scouts*. Unpublished manuscript in author's collection, 1944.

McConnell, Zeke. *Diary: Alamo Scouts*. Unpublished, 1944.

Niles, Gibson. *The Operations of the Alamo Scouts (Sixth U.S. Army Special Reconnaissance Unit)*. Fort Benning: U.S. Army Infantry School, 1948.

Ross, Bob. *Diary: Alamo Scouts, Sixth Army*. Unpublished, 1945.

Rounsaville, Tom J. *The Operations of the Alamo Scouts (Sixth Army's Reconnaissance Unit)*. Fort Benning: U.S. Army Infantry School, 1950.

Zedric, Lance Q. "Eyes Behind the Lines: Sixth Army's Special Reconnaissance Unit of World War II." Master's dissertation, Western Illinois University, 1993.

Zedric, Lance Q. *Silent Warriors of World War II: The Alamo Scouts Behind Japanese Lines*. Ventura, CA: Pathfinder Publishing, 1995.

Zedric, Lance Q., and Michael F. Dilley. *Elite Warriors: 300 Years of America's Best Fighting Troops*. Ventura, CA: Pathfinder Publishing, 1996.

Articles

Author unknown. "Alamo Scouts, Sixth Army." Source and date unknown.

Author unknown. "Enemy on Luzon: An Intelligence Summary (Chapter IV, Special Reconnaissance Operations)." Source and date unknown.

Author unknown. "The Saga of Bill Nellist—Alamo Scout." *Airborne Quarterly*, Spring 1991.

Chronis, Peter G. "Alamo Scouts: Masters of Stealth." *Denver Post*, October 5, 1993.

Dilley, Michael F., and Lance Q. Zedric. "The Recon of Los Negros." *Behind The Lines*, May–June 1995.

Hughes, Les. "The Alamo Scouts." *Trading Post*, April–June 1986.

Johnson, Raymond, and Alfred Hahm. "Alamo Scouts, U.S. Sixth Army: 1943–1945." *Company of Military Historians*, Plate 499.

Lindsey, Beverly. "CSM Galen C. Kittleson Retires: Last of the Alamo Scouts." *Static Line*, September 1978.

Nabbie, Eustace C. (pseudonym of Mayo S. Stuntz). "The Alamo Scouts." *Studies in Intelligence*, date unknown.

Pomes, George. "The Great Cabanatuan Raid." *Air Classics*, two issues, date unknown but probably between 1981 and 1984.
Raymond, Allen. "Team of Heroes: The Alamo Scouts." *The Saturday Evening Post*, June 30, 1945.
Shelton, George R. "The Alamo Scouts." *Armor*, September–October 1982.
Spencer, Murlin. "Saga of the Alamo Scouts." *Detroit Free Press*, 15 October 1944.
Wells, Jr., Billy E. "The Alamo Scouts: Lessons for LRSUs." *Infantry*, May–June 1989.
Zedric, Lance Q. "Prelude to Victory: The Alamo Scouts." *Army*, June 1994.
Zedric, Lance Q., and Michael F. Dilley. "Raid on Oransbari." *Behind the Lines*, November–December 1995.

Interviews and Letters

Kittleson, Galen C. Several interviews between 1992 and 1999.
Lupyak, Joseph W., SGM (ret.). Telephone interview, September 30, 1996.
Nellist, William. Several interviews between 1992 and 1995.
Rounsaville, Tom J. Several interviews in 1992, 1995, and 1998.
Santos, Terry. Several interviews and letters between 1992 and 2001.
Smith, Andy. Several interviews in 1992, 1995, and 1998.
Stuntz, Mayo S. Several interviews in 1992, 1995, and 1998.
Sumner, Robert S. Several interviews and letters between 1992 and 1999.

CHAPTER 7

Special Allied Airborne Reconnaissance Force

In early 1945, with the Allied forces closing in on Germany from the west and Soviet forces from the east, there was concern on the Supreme Headquarters Allied Expeditionary Force (SHAEF) staff about the continued safety of POWs in the hands of the Germans. Some intelligence reporting indicated that the Germans were possibly moving POWs out of camps in areas where Allied military forces were near at hand. Fears of maltreatment of POWs in the event that Germany lost the war had been raised earlier but planning for those contingencies did not receive a high priority.

As the situation became more of a reality and concerns were raised that German camp personnel might decide to massacre prisoners, the priority for planning was raised. The new plans centered on establishing "Contact Teams" which could be parachuted into or near POW camps to observe and report activities relative to prisoner safety or to intercede in the event of any observed untoward actions being directed against prisoners. Based on limited intelligence reporting, conditions in some of the camps were believed to be bad. There was also uncertainty about whether Hitler might order that actions be taken against prisoners in view of Germany steadily losing ground in the war. One scenario imagined prisoners being killed; another that the camps would simply be abandoned and the prisoners might starve or otherwise be left to fend for themselves. The renewed planning called for parachute battalions to be held on alert in reserve, prepared to jump in on camps if any contact team notified them of imminence of killings or mass evacuation of prisoners. Initially, the 501st Parachute Infantry Regiment of the U.S. 101st Airborne Division was designated as the unit on stand-by.

In early March 1945, SHAEF created a multi-national organization to oversee and coordinate the activity of the contact teams. Initially, this organization was to be manned by volunteers from among British, American, and Belgian paratroopers who would be part of a "dangerous undercover mission" which was originally designated Operation *Vicarage*. This organization was re-designated as the Special Allied Airborne Reconnaissance Force (SAARF) by its new leader. A British officer, Brigadier J. S. Nichols, was appointed as commander and an American, Colonel J. E. Raymond, as deputy commander.

The basic contact team was to be composed of three men. Team composition was to consist of two officers and an enlisted radio operator. Unlike some previous organizations, specifically the OSS/SOE Jedburghs, all members of each team were to be of the same nationality. A high priority was placed on finding team members with German or Eastern European language capability. In all, the SAARF consisted of 120 French, 96 British, 96 American, 30 Belgian, and 18 Polish soldiers. Among the Americans, the volunteers were drawn from OSS units (including some former Jedburghs), the 82nd and 101st Airborne Divisions, a small group from the 13th Airborne Division, and one U.S. Navy radio operator who had previously been part of the OSS. The British, Belgian, and French soldiers came from several special operations forces (including SOE and SAS brigades), while the Poles were drawn from the Polish Independent Grenadier Company.

Several of the volunteers were women from the SOE who had parachuted into France and served with the French Resistance. The SAARF planners decided that, despite the impressive records the women in SAARF made in their previous assignments, they would not be assigned to any of the airborne teams. This did not preclude them being deployed on the air transportable or jeep mounted teams. Needless to say, the women were disappointed.

Those volunteers who were not already airborne qualified were sent to the No. 1 Parachute Training School at Ringway, England, for training and qualification. Training was to be conducted by the SOE, with assistance to be provided by the OSS. The training objective was to have 60 teams ready to be dispatched by 1 May 1945. Altogether, the plan called for 120 teams of three people each. In order to facilitate training as well as command and control of the teams, SAARF was organized into three contingents. The 1st Contingent comprised British and Belgian teams; the 2nd Contingent included American and Polish teams; and the 3rd Contingent was made up of French teams.

Headquarters for SAARF was located at the Sunningdale golf course at Wentworth in Surrey, England, which had formerly been used by the 21st Amy Group. OSS and SOE were tasked to provide all training and support personnel. The First Allied Airborne Army supplied the operational personnel. Command and control of SAARF remained with SHAEF.

While the contact teams were in training, SHAEF prepared leaflets in German to be dropped in front of the advancing Allied units. These leaflets warned German units that any captured SAARF personnel were not to be considered as terrorists but on a humanitarian mission. A strong warning was contained in the leaflets that any SAARF personnel executed by German forces would be considered a war crime and dealt with harshly following the end of hostilities. The leaflets were dropped in late March.

As the team training progressed and general plans for their deployment were made, the progress of the war impacted the planning. On 24 March, in a move predicted the day before by Axis Sally on Radio Berlin, Operation *Varsity*, the airborne leap across the Rhine, kicked off with parachute and glider assaults, moving Allied forces into Germany in force and even closer to Berlin.

In mid-April, SAARF received a request for a mission to be deployed in the vicinity of the prison camps in and around Altengrabow. Planning went into high gear, but the mission was almost immediately rejected. On 21 April, the Belgian government requested that eight of its 10 contact teams be sent to Brussels. There the teams were to be used in a ground role by various army groups to collect information available concerning conditions in POW camps. Almost as soon as these teams were dispatched, SAARF reorganized. The 60 contact teams that had already completed their training were designated for airborne missions, while the others were designated for either air or ground transportable missions.

On 23 April, an urgent request was made to SAARF to send in six contact teams to the Altengrabow camps, the mission that had earlier been rejected by SAARF. This request was based on information that prisoners were being moved further east and may possibly be mistreated by camp personnel. The previous plan was reinstated and contact teams were put on alert for deployment in two days. The contact teams were placed under the command of British Major Phillip Worrall, an SOE asset from the South Wales Borderers.

This mission was given the code name of *Violet*. The primary mission target was Stalag XIA and other camps in the Altengrabow area. These camps were located between the Western Allied divisions pressing in from the west and Soviet Red Army units closing in from the east. The six contact teams, whose members would meet for the first time at the mission briefing, and their make-up, were as follows:

Team	Team Leader	Description
Drop Zone 1:		
ERASER	Major Phillip Worrall	British Team
BRIEFCASE	*Sous-lieutenant* (S/Lt) Cousin	French Team
Drop Zone 2:		
PENNIB	Major Sam Forshall	British Team
CASHBOX	Captain J. Brown	American (OSS) Team (included one officer and two radio operators)
Drop Zone 3:		
SEALINGWAX	*Capitaine* Paul Aussaresses	French Team
PENCIL	Captain Warfield	American (504th PIR/OSS) Team

During the mission briefing, the various contact team members were each given a letter of authority, which identified them as SAARF personnel. The letters were printed in English on one side and German on the other side. Each team member was identified in his letter by name, rank, and serial number. The purpose of the mission was described as to determine if Allied POWs were being treated in accordance with the Geneva Convention and requested that any assistance from camp personnel be given upon request. The letters were signed by Brigadier J. S. Nichols, the SAARF commander, and General T. J. Davis, an American.

On 25 April, at 2030, three planes lifted off from their departure airfield at RAF Great Dunmow. Operation *Violet* had begun. In a not unusual airborne drop, the contact teams were scattered in the air, and consequently on the ground. No immediate attempt was made by the six teams to contact any of the camps. Within a few days, ERASER team was captured by the Germans and moved immediately to Stalag XIA. Soon after, S/Lt Cousin of BRIEFCASE team and the OSS radio operators of CASHBOX team joined them in the prison camp.

After some discussions with the Stalag XIA commandant, Colonel Ochernal, Major Worrall convinced him that Western Allied and Russian (Soviet) forces were closing in on his camp and urged him to cooperate with the SAARF members. Ochernal agreed to allow the SAARF contact team members already in the camp, under the close observation and supervision of German personnel, to review conditions in the camp and even to transmit messages back to SAARF Headquarters. In a surprise move, Ochernal produced a captured SOE radio for Worrall to use.

Worrall reported back on the members of the SAARF who were present in the camp and that they were being treated by the Germans with a limited amount of courtesy, although there was still a good deal of tension within the camp. He went on to report that there were fewer prisoners in the general area than had been anticipated. Expectations prior to the operation kicking off were that there would be as many as 100,000 prisoners in the area. It turned out that there were only about 20,000 POWs there; about 2,000 of them were Americans and British as well as some French and Belgians.

On 2 May, SAARF Headquarters radioed Worrall to inform him that an agreement had been reached that provided that the nearby American 83rd Infantry Division would supply trucks to begin the immediate evacuation of Allied POWs. The 83rd Division had, in mid-April, liberated a subordinate camp of the notorious Buchenwald concentration camp, located at Langenstein. Ochernal agreed to provide security details for truck convoys. The next day, 70 trucks and 30 ambulances arrived at Altengrabow. There were 40 war correspondents in the convoy who had come to witness the prisoners' liberation. The evacuation began immediately. This continued until the afternoon of 4 May. At that time, forces from the Red Army arrived at the camp and took control. The mood in the camp changed with the arrival of the Red Army, whose soldiers permitted the remaining Western prisoners to depart but would not allow the evacuation of any of the Poles or Italians, even though these prisoners had requested repatriation to the West.

In the morning of 5 May, Worrall was told by the senior Red Army officer that, despite Worrall's objections, the SAARF team members in the camp had two hours to gather their gear and return to their own lines, and that the Red Army would handle all further matters concerning the camp and prisoners from that point on. For all intents and purposes, Operation *Violet* was concluded. The operation did not proceed quite as planned but it was, in the long run, successful in securing the release of all the Western Allied prisoners. None of the SAARF contact team members were killed or injured.

Meanwhile, the SEALINGWAX team had been dropped some 15 miles away from its intended drop zone. They were dropped into an area controlled by a German division, but they were able to elude capture by the Germans. On 15 May, however, they were captured by the Red Army while wearing German SS uniforms near the town of Kuhberge. After three days of intensive interrogation by the Russians, the team members were moved to a Soviet POW camp at Zerbst because their captors believed that this French team was lying about what its mission was. Once in the Soviet camp, the SAARF team members were stripped of their uniforms, possessions, and documents. They were interrogated, especially about the SAARF mission. On 7 June, the team members escaped and evaded successfully through Americans lines at Halle and eventually made it back to SAARF Headquarters, arriving just as the organization was being disbanded.

Operation *Violet* was the only airborne insertion of any SAARF contact team. However, 74 other teams were inserted by jeep or air to their various missions across Northern Europe during the remainder of the war. These teams conducted such missions as working with local military governments to set up communications links, translating captured documents, interrogating captive German prisoners, and monitoring the movement of German forces as they returned to Germany. In addition, the teams screened the German prisoner population to segregate those determined to be political prisoners, criminals, and those Nazis who would later be charged with war crimes.

One member of a SAARF contact team located in the vicinity of Lake Constance had an unfortunate run-in with a notorious British deserter and collaborator. Captain Frank Lillyman, of the U.S. 101st Airborne Division, was approached by the British deserter, Harold Cole. Cole had managed to get away from Paris, where he had worked with the Gestapo during much of the war rooting out downed Allied air crew personnel and escaped prisoners. Once in Allied areas of control, Cole persuaded a U.S. Army Counter-Intelligence Corps (CIC) detachment that he was an escaped British prisoner. The CIC unit provided Cole with identification papers and an American army uniform and turned him over to Lillyman to assist in the search for war criminals.

Cole convinced Lillyman that a local Frenchman, who owned a Mercedes, was a traitor and should be placed under arrest. When Lillyman went to arrest the Frenchman, Cole shot the Frenchman dead and stole his car. Cole immediately reported the incident to local American forces but told them that Lillyman had shot and killed the Frenchman. Although Lillyman was arrested and held for court-martial, the truth was eventually uncovered and he was released, with no legal action taken against him. Cole was later killed in Paris in a shoot-out with local police officers.

Another contact team, including Major Henry Coombe-Tenant (of the Welsh Guards) and another British officer, Patrick Leigh Fermor (from SOE by way of the Irish Guards), were given this mission to learn as much as they could about the impregnable fortress known as Oflag IVC, or Colditz Castle, in Saxony. Several prominent British prisoners were kept at Colditz, including Lieutenant Colonel David Stirling (founder of the British SAS) and Lord Harewood (a cousin of the king), as well as relatives of Winston Churchill and Field Marshal Harold Alexander of Tunis fame. Fermor learned that a former prisoner who had been released in a prisoner exchange from Colditz because of health reasons, Miles Reed of the Phantom Reconnaissance Force, was living at his home in Haslemere. Fermor rushed to talk with Reed to learn as much as he could about the area surrounding Colditz. When Fermor told Reed of the SAARF plan (which involved the SAARF contact team slipping into the castle dressed in tattered POW uniforms, ostensibly as members of a prisoner working party), Reed exploded and told him that the plan had no chance of succeeding. In the first place, Reed said, there were no working parties because all of the prisoners were officers. Reed was so certain that the plan would fail that he threatened to go to Churchill if the plan was not cancelled. His visit was not necessary, however, as the thrust of General George S. Patton's Third Army arrived at the castle on 16 April and freed the prisoners.

On 1 July, having completed its unusual but humanitarian mission, SAARF was disbanded. Its contact teams members were returned to their parent units.

Members of SAARF were issued two specific insignia. The first insignia issued was the SAARF title. This was a cloth rectangle of dark blue material with the organization initials in yellow, each letter followed by a period. This insignia was intended to be worn on the left sleeve of the uniforms of Allied soldiers, beneath the SHAEF shoulder patch and beneath the unit shoulder patches of American soldiers. Generally speaking, since American soldiers were not used to title insignia, they did not wear this unit insignia.

The other insignia was a unique wing design, intended to be worn on the right shoulder of the uniform. Allied soldiers wore the SAARF wing there. American uniforms customarily had the individual soldier's combat unit patch worn on the right shoulder; consequently, most Americans generally preferred to wear the SAARF wing near the cuff of their right sleeve. The SAARF wing was silver-blue on royal blue wool. The wing ended in a red arrow which appeared to break chains, symbolizing the organization's mission of aiding in the release of Allied prisoners of war.

Sources

Books
Breuer, William B. *Geronimo! American Paratroopers in WWII*. New York: St. Martin Press, 1989.
Critchell, Laurence. *Four Stars in Hell*. New York: MacMillan, 1947.
Nichols, John, and Tony Rennel. *The Last Escape: The Untold Story of Allied Prisoners of War in Germany, 1944–1945*. United Kingdom: Penguin Books, 2003.
Smith, Bradley F. *The Shadow Warriors: O.S.S. and the Origins of the C.I.A*. New York: Basic Books, 1983.
Whittaker, Len. *Some Talk of Private Armies*. London: Albanium Publishers, 1984.

Articles
Downing, Ben. "A Visit with Patrick Leigh Fermor, Part 2." *The Paris Review Daily*, 24 May 2013.
Hughes, Les. "The Special Allied Airborne Reconnaissance Force (SAARF)." *Trading Post*, issue unknown, 1991.
"Obituary—General Paul Aussaresses." *London Telegraph*, 4 December 2013.

Internet Sources
Author unknown. "Special Allied Airborne Reconnaissance Force." Wikipedia, no date.
Capp, Jimmy. "SAARF, Special Allied Airborne Reconnaissance Force." Posted 14 June 2009. https://www.usmilitariaforum.com.
Dilley, Michael F. "A Short History of U.S. Airborne Units in World War II." *Tidbits*, 21 August 2012.
Durand. "SAARF, POWs, and Tensions between Allies." Posted 5 June 2003. https://forum.axishistory.com.
Jedburgh22. "SAARF." Posted 31 October 2010. https://www.ww2talk.com/index.php.

Note: All insignia/pocket patches are from the author's own collection and Wikimedia.

Shoulder sleeve insignia of the major airborne units that operated in Europe. Clockwise from top left: 1st Allied Airborne Army, Glider patch for overseas cap showing wearer was on jump status, 101st Airborne Division, 17th Airborne Division, 13th Airborne Division, 82d Airborne Division, and (center) XVIII Airborne Corps.

Shoulder sleeve insignia of the major airborne units that operated in the Pacific. Left to right: 11th Airborne Division and the 503d Parachute Infantry Regiment.

Shoulder sleeve insignia of the ghost airborne divisions. By row, left to right: 6th Airborne Division, 9th Airborne Division, 18th Airborne Division, 21st Airborne Division, and the 135th Airborne Division.

The pocket patch of the 501st Parachute Infantry Regiment.

The distinctive insignia of the 501st Parachute Infantry Regiment.

The distinctive insignia of the 506th Parachute Infantry Regiment.

The shoulder sleeve insignia of the First Special Service Force.

Collar branch insignia worn by U.S. and Canadian officers of the First Special Service Force.

Parachute insignia, as worn by those who successfully completed airborne training or who participated in a combat jump without having been previously qualified.

Glider badge, as worn by those who successfully completed glider operations training or who participated in a combat operation without having been previously qualified.

Navy and Marine Corps jump wings, as worn by those who had completed airborne training and had made an additional five jumps.

Replica of unauthorized shoulder patch of the Alamo Scouts. This patch was based on the winning design by Corporal Harry A. Golden, a medic at the Alamo Scouts Training Center. The contest, held in 1944, was to find a design for a shoulder patch. The design was submitted to the A.N. Meyer Company, which made 440 patches. The patch was never authorized by the U.S. Army but was worn by Scouts on their left shoulder during the war and on the right shoulder of those who remained in the army. (Photo courtesy of the Alamo Scouts Association. Used with permission.)

SAARF wing, worn by non-Americans of the unit on the right shoulder of their uniform sleeve. American members of the unit preferred to wear it near the right cuff, similar to the Pathfinder wing.

SAARF shoulder title, worn by non-American members of the unit on the left sleeve, below the wing. Because of the general unfamiliarity with American titles, they usually did not wear it.

Shoulder sleeve insignia of the 11th Airborne Division.

Pocket patch awarded to graduates of the MACV Recondo School.

Some examples of the various unit Recondo insignia. Left to right, by row: 101st Airborne Division, 82d Airborne Division, 9th Infantry Division, 25th Infantry Division, 5th Infantry Division, and the XVIII Airborne Corps.

CHAPTER 8

Gypsy Task Force at Aparri

While airborne operations involving insertion by parachute were conducted periodically in the Southwest Pacific Theater during World War II, gliders were used only once in a tactical operation. Although four gliders had been used to ferry in engineers and their equipment at Nadzab, New Guinea, in September 1943, this was not a tactical use. The engineers were sent in to construct an emergency air strip several days following its capture by the 503rd Parachute Infantry Regiment. The only tactical use of gliders occurred in the Philippines near the end of the fighting there.

By the end of April 1945, much of the Japanese resistance in the southern part of Luzon, the largest island in the Philippine chain, had ended. This was marked by the capture of Mount Malepunyo, near the Japanese airfield at Lipa, and the virtual destruction of the *Shimbu* Group of the Japanese Fourteenth Army. This was accomplished by the 11th Airborne Division, under the command of Major General Joseph Swing. During that operation, the 11th had been acting in conjunction with the 1st Cavalry Division and elements of Philippine guerrilla units in clearing southern Luzon of the Japanese forces who fought until the very last man, often accepting suicide rather than surrender. It was also during this period that elements of the 11th Airborne Division conducted their successful raid on the Japanese POW camp at Los Baños.

However, Japanese forces under the command of General Tomoyuki Yamashita (the "Tiger of Malaya," who had conquered the British in the Malay Peninsula in 70 days) continued to fight in the north following the defeats of other groups of the Fourteenth Army in southern Luzon. These troops were the *Shobu* Group, the largest group of the Fourteenth Army, numbering about 152,000. They concentrated their fighting in the Sierra Madre and Cordillera Central mountains areas.

General Walter Krueger, commander of the U.S. Sixth Army, ordered his staff to develop a plan for conquering the northern part of Luzon as quickly as possible. One aspect of this plan was to block Japanese forces from escaping from the Philippines through the northern port of Aparri. Aparri was a small river port located at the mouth of the Cagayan River. Krueger was determined to deny the Japanese the use of this port.

Early planning considered the possible use of the 11th Airborne Division. This division was in the process of training replacements at Lipa, 40 miles south of Manila. While there, the paratroopers learned that the war had ended in Europe with the surrender of Germany. Within a week, after having been alerted by Krueger's staff of the possibility of a new mission that would involve parachute insertion, the division began jump training at the airfield. This training was designed to refresh older members of the division who had not jumped recently and to qualify additional members. These new members included glider troopers from the 187th and 188th Glider Infantry Regiments of the 11th who were not already jump qualified. The 54th Troop Carrier Wing and its aircraft arrived at the airfield to conduct the actual jumps for the division. During May, the 54th carried paratroopers of the 11th on five training jumps.

In addition to the possible airborne operation in northern Luzon, the 11th was also hearing rumors about a possible combat jump into China, where stubborn Japanese resistance was still being encountered, and even parachute and glider landings on the main island in a potential invasion of Japan. All of these rumors made the training that much more urgent.

By now, the 37th Infantry Division (an Ohio National Guard unit) had pushed Japanese forces from their positions in the Cagayan Valley and was pursuing them as they moved northward along Route 5. The 37th was assisted in this pursuit by the 6th Infantry Division and elements of Philippine guerrilla units. Following behind the 37th and the 6th was the 25th Infantry Division, with the 126th Infantry Regiment (of the 32nd Infantry Division) attached.

The more that Krueger and his staff studied the tactical situation as it was developing, the more convinced they became that the port of Aparri had become important to the plans of the Japanese force. Krueger decided that it was necessary to use the 11th Airborne Division to block this potential escape route. His staff then notified the 11th of firm plans that it was to conduct an airborne operation at Aparri, and once on the ground was to move in a southerly direction along Route 5 to conduct operations in conjunction with the 37th Infantry Division that was coming up from the south. In order to conduct this mission, the staff created two task forces, Task Force Connolly and Gypsy Task Force.

Task Force Connolly was composed of about 800 soldiers, including a reinforced company of the 127th Infantry Regiment (32nd Infantry Division), a company of the 6th Ranger Battalion, a battery of 105mm howitzers, and a group of detachments composed of engineer, medical, and port soldiers. The leader of the task force was Major Robert V. Connolly of the 123rd Infantry Regiment (of the 33rd Infantry Division). Its mission was to move from Vigan, where it was organized, north along Route 3 and around the northwest tip of Luzon, and to meet up with a battalion of the Philippine 11th Infantry, a guerrilla unit, at Aparri. Once there, the two groups would secure and prepare the drop zone for the paratroopers and gliders of the 11th Airborne Division.

Gypsy Task Force was composed of three companies of the 1st Battalion, 511th, two companies of the 3rd Battalion, 511th, and supporting units. These supporting units included: Battery C, 457th Parachute Field Artillery Battalion; the 1st Platoon of the 127th Airborne Engineer Battalion; the Demolition Platoon of Headquarters Company 511th; the 2nd Platoon of the 221st Airborne Medical Company; a detachment from Service Company, 511th; and a detachment from Special Troops, 511th that consisted of the 511th Parachute Maintenance Company, 511th Airborne Signal Company, and the 711th Airborne Maintenance Company. The Gypsy Task Force consisted of 989 soldiers, and was commanded by Lieutenant Colonel Henry Burgess, commander of the 1st Battalion, 511th.

Burgess' mission was to bring his task force in to Camalaniugan Airfield, which had been abandoned and was now overgrown. This airfield was

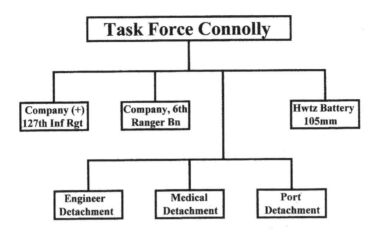

Organization of Task Force Connolly, part of the 11th Airborne Division's Gypsy Task Force. (Chart by author)

located along the Cagayan River five miles south of Aparri. Fifty-four C-47 Dakotas, 14 C-46 Commandos, and seven gliders were put at Burgess' disposal for this operation. The gliders were six Waco CG-4As and one Waco CG-13A. The equipment in the gliders consisted of 19 trucks, six Jeeps, a trailer, and supplies. The aircraft were from the 317th Troop Carrier Group and the 433rd Troop Carrier Group. The aircraft were commanded by Colonel John Lackey.

Intelligence reports developed during the planning phase showed that the *Shobu* Group had shifted its operations to the Cordillera Central Mountains, west of the Cagayan Valley. As a result of this shifting, the 6th Infantry Division ran into heavier fighting and the 37th Infantry Division was able to advance fairly quickly north along the Cagayan River, facing lighter resistance. The target date for Gypsy Task Force to jump in had been set for 25 June; it was now moved up to 23 June.

By the 21st, the soldiers of Connolly Task Force had already occupied the airfield at Aparri, and by the 22nd ensured that there was no enemy resistance in the area. Despite recommendations by his staff that he cancel the airborne operation, Krueger insisted that it be conducted as planned. Once this decision was made firm, word was passed to Connolly Task Force and its soldiers, acting as pathfinders, to be prepared to send out smoke signals to mark the drop zone as the aircraft approached.

GYPSY TASK FORCE AT APARRI • 85

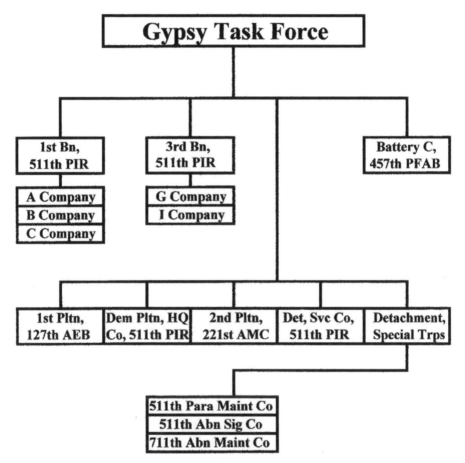

Organization chart of the Gypsy Task Force of the 11th Airborne Division. (Chart by author)

Gypsy Task Force paratroopers assemble on the drop zone. (The American Warrior website)

Gypsy Task Force jumping onto drop zone at Aparri. (Library of Congress)

On the morning of the 23rd, at Lipa airfield, the men of Gypsy Task Force loaded their aircraft by 0430. By about 0600 the first aircraft rolled down the runway and took off. By 0900 fighter aircraft from the Fifth Air Force had already laid down a smoke screen over the drop zone area and the hills located southeast of the airfield, and cleared out of the area. The pathfinders on the ground began sending up their colored smoke signals when they were alerted that incoming aircraft were in sight. With the C-47s in the lead, followed by the C-46s and, finally, the gliders in tow, the aircraft approached the drop zone.

The C-47s came in higher than the planned altitude of about 700 feet. When the pilots of the C-46s noticed this, they climbed their aircraft to about 1,100 feet in order to avoid any descending paratroopers in front of them. Winds on the drop zone were running at about 25 miles per hour, higher than anticipated. By now all of the paratroopers were hooked up and standing by for the green light to come on. Despite the high winds, the green light did come on, clearing the paratroopers to jump. The paratroopers quickly exited their aircraft. Although the high winds were a factor, all of the paratroopers landed on the drop zone, as did all seven gliders, although two of the gliders suffered crash landings. Despite the successful landings, casualties were higher than expected due to the winds and deep ruts on the drop zone: two paratroopers were killed and about 70 were injured in the drop.

Once all the paratroopers were on the ground and the gliders were unloaded, Burgess ordered an accounting of all his troops. When he was satisfied that everyone was assembled and ready, and that the casualties were being taken care of, he gave the order to move out along Route 5. Patrols were sent out in advance to scout the area and to alert the rest of the task force of any enemy encountered. Soon the paratroopers encountered heavy Japanese resistance. At this point, they used flame throwers to clear caves, bunkers, and pockets of soldiers. Heavy fighting continued for three days as Gypsy Task Force continued to move south along Route 5.

Finally, on 26 June, the paratroopers of Gypsy Task Force and lead elements of the 37th Infantry Division hooked up together, completing the encirclement of the *Shobu* Group. This effectively marked the strategic end of the campaign for northern Luzon. Fighting continued to mop up the remnants of Yamashita's forces but the men of Gypsy Task Force were ordered to return to Aparri. There they linked up with troop carrier planes that returned them to Lipa, having successfully completed their mission.

Sixth Army estimated that only about 23,000 soldiers remained of *Shobu* Group, and that they were sufficiently disorganized so as to not constitute an effective resistance. Following the removal of Gypsy Task Force, MacArthur turned control of the fighting in northern Luzon over to Eighth Army. In fact, the estimates of the strength of Japanese forces remaining in northern Luzon were off by more than almost 30,000, and it would take Eighth Army more than a month-and-a-half to bring the fighting to a halt.

Gypsy Task Force was moved out of the northern Luzon area in order to ready Sixth Army, and specifically the entire 11th Airborne Division, to begin its preparations for the invasion of Japan. By now the invasion plans had changed and the combat jump had been cancelled. The 11th was to be part of the reserve forces in that invasion.

Soon after its return, the 11th began to receive more reinforcements to replace its almost 1,900 casualties in the Philippine campaign. This reinforcement consisted primarily of the newly arrived 541st Parachute Infantry Regiment from Fort Benning, under the command of Colonel Ducat McEntee. Earlier in the war, men of the 541st had been reassigned to various airborne units on an individual basis. The regiment was now prepared to take its place, as a unit, in the war. But they were disappointed once again. A theater decision was made to disband the 541st and send its members to the various units within the 11th.

In the meantime, following the atomic blasts at Hiroshima and Nagasaki, Japan sued for peace to bring the fighting in the Pacific and East Asia to an end. The 11th Airborne Division and 1st Cavalry Division moved to Okinawa on 12 August to prepare for the occupation of Japan. The jump at Aparri was the last airborne operation conducted in World War II.

There are some critics who disparage airborne units jumping into areas where the drop zones have been secured by ground forces already on the scene, as it was at Aparri. However, in almost every case (the 504th and 505th Parachute Infantry Regiments at Salerno, the 503rd Parachute Infantry Regiment at Noemfoor Island, Gypsy Task Force at Aparri, the 173rd Airborne Brigade jump in Tay Ninh Province during the war in Vietnam, and even the 173rd jump in Bashur during the war in Iraq, to name just a few), the need for fast delivery of soldiers to the battlefield and the results these paratroopers gathered thereafter have proven the value of such operations.

Sources

Books

Bradley, Francis X., and H. Glen Wood. *Paratrooper*. Harrisburg, PA: Stackpole Books, 1962.

Breuer, William B. *Geronimo! American Paratroopers in World War II*. New York: St. Martin's Press, 1989.

Devlin, Gerard M. *Paratrooper! The Saga of U.S. Army and Marine Parachute and Glider Combat Troops During World War II*. New York: St. Martin's Press, 1979.

Devlin, Gerard M. *Silent Wings: The Saga of the U.S. Army and Marine Combat Glider Pilots of World War II*. New York: St. Martin's Press, 1985.

Flanagan, Jr., Edward M. *11th Airborne*. Paducah, KY: Turner Publishing, 1993.

Flanagan, Jr., Edward M. *Airborne: A Combat History of American Airborne Forces*. New York: Ballantine Publishing, 2002.

Flanagan, Jr., Edward M. *The Angels: A History of the 11th Airborne Division 1943–1946*. Washington, D.C.: Infantry Journal Press, 1948.

Flanagan, Jr., Edward M. *The Angels: A History of the 11th Airborne Division*. Novato, CA: Presidio Press, 1989.

Galvin, John R. *Air Assault: The Development of Airmobile Warfare*. New York: Hawthorne Books, 1969.

Huston, James A. *Out of the Blue: U.S. Army Airborne Operations in World War II*. West Lafayette, IN: Purdue University Studies, 1972.

Krueger, Walter. *From Down Under to Nippon: The Story of Sixth Army in World War II*. Nashville: Battery Classics, 1989.

Lassen, Don, and Richard K. Schrader. *Pride of America: An Illustrated History of the U.S. Army Airborne Forces*. Missoula, MT: Pictorial Histories Publishing, 1991.

Mrazek, James E. *The Glider War*. London: Robert Hale, 1975.

Rottman, Gordon L. *US Army Airborne 1940–90: The First Fifty Years*. London: Osprey Publishing, 1990.

Smith, Robert Ross. *United States Army in World War II. The War in the Pacific: Triumph in the Philippines*. Washington, D.C.: U.S. Army Center of Military History, 1963.

Steinberg, Rafael. *Return to the Philippines*. Alexandria, VA: Time-Life Books, 1979.

Articles

Bruning, John H. "The Last Jump: Task Force Gypsy at Aparri." Posted in 2014. https://theamericanwarrior.com/2014/12/05/the-last-jump-task-force-gypsy-at-aparri/.

CHAPTER 9

Recondo Training—Its Origins and Aftermath

Many folks are not aware that the concept for Recondo training began during World War II in the Southwest Pacific Theater. It began at the Alamo Scouts Training Center (ASTC), an organization that trained members of the Special Reconnaissance Unit of Sixth Army, known as the Alamo Scouts. The idea for the ASTC began with General Douglas MacArthur, although he didn't realize it. He would not permit organizations to operate in his theater that did not report to him. Thus, since the OSS was commanded from its headquarters in Washington, D.C., and reported to that headquarters, MacArthur would not allow OSS to operate in the Southwest Pacific. However, realizing that he needed an organization similar to OSS to oversee intelligence gathering, as well as sabotage and other operations, he created, under the supervision of his Assistant Chief of Staff G-2, the Allied Intelligence Bureau.

The idea of initiating the organization of units to support a command was not lost on MacArthur's subordinate commanders. In November 1943, taking his cue from MacArthur, General Walter Krueger, commanding general of Sixth Army, created the ASTC to train small teams that would conduct intelligence gathering, reconnaissance, raiding, and indigenous unit training in guerrilla operations within the Sixth Army's area of operations. The training lasted for six and one-half weeks, including four weeks of training and two and one-half weeks of field training exercises to implement the previous weeks of training.

The training curriculum for the first four weeks consisted of two hours of physical training and swimming every day. Other subjects included: cover and concealment, unarmed combat (Judo), use of rubber boats, map reading, scouting and patrolling, stalking, movement under observation and enemy fire, familiarization with Japanese weapons, range firing, night scouting, boat reconnaissance, observation, combat intelligence, sketching shore line and beach installations, jungle food sources, intelligence reports, night land navigation, ropes and snares, aerial photography, explosives, knife fighting, and booby traps.

In many aspects, this training was similar to the training the original men of Colonel William Darby's 1st Ranger Battalion received at the hands of the British Commandos at their training sites in Scotland in 1942. That stressed physical training, with an emphasis on speed marches with full equipment, day after day. Other training included river crossings, mountain and cliff climbing, sliding down ropes over water, and swimming, especially in ice cold water. There was also training with friendly and enemy weapons, hand-to-hand fighting, scouting, patrolling, small unit tactics, map reading, and first aid. Emphasis was placed on nighttime exercises and amphibious landings from various crafts. Almost all amphibious exercises were conducted under live-fire conditions. As one writer said, "The speed march was the heart of the Ranger toughening process ..." The Rangers "grew accustomed to marching through fatigue and pain."

At the ASTC training, teams of six or seven men were selected to make up the Alamo Scouts basic operating units. Krueger recognized that this would mean that not all of those who were trained and graduated from the ASTC would make up the teams. Krueger instructed his subordinate units, who had provided the volunteers who were trained at the ASTC, that this did not mean, and would not be construed to mean, that those who were returned to their unit were any less qualified than those who made up the teams. Those men who graduated from the ASTC and returned to their units were to pass on to other members of their units what they had learned while at ASTC. All told, 400 soldiers attended the ASTC during the nine courses that were conducted between December 1943 and September 1945. Of that number, 117 enlisted men and

21 officers joined the various Alamo Scouts teams. A total of 19 teams were formed, although never more than 10 were active at any one time.

The concept of having graduates return to their units to pass on their training to members of those units was kept alive when in September 1950, the Infantry School at Fort Benning, Georgia, created the Ranger School. This school was started to train soldiers who would make up the various Ranger/Airborne companies that were being raised by the army to be employed in Korea and elsewhere. Ranger School lasted about eight weeks and was conducted in several locations, including a mountainous area in Georgia and in a Florida swampy area. In a throwback to its training roots by the Commandos, the course retained the rope slide over water. When this training course was established, the Infantry School recognized that not everyone who graduated from Ranger training would be assigned to the companies. Thus, the concept of returning Ranger-qualified soldiers to their units to pass on their training was instituted and is still active to this day. Following the end of the Korean War, the Ranger/Airborne companies were disbanded but the Ranger School continued.

In 1956 in Augsburg, Germany, the 11th Airborne Division created the first Long Range Reconnaissance Patrol concept. Because of the area the division patrolled, it was thought necessary to have a force that could operate deep into East Germany or Czechoslovakia in the event of war. From the 11th Airborne, the need to establish additional LRRP companies expanded throughout the Seventh Army in Germany. The individual patrols consisted of five or six men, similar in organization to the Alamo Scouts. At the time, there was no formal training established for the volunteers who made up the strength of the companies. However, many of those in leadership positions, both officers and non-commissioned officers, were combat veterans who had served during World War II or the Korean War or who had Ranger or special forces training and were able to teach their men based on their experiences and previous training. Training concepts concentrated on communications, techniques of patrolling, raiding, reconnaissance, surveillance, and insertion and extraction methods.

In late 1958 at Fort Campbell, Kentucky, General William Westmoreland commanded the 101st Airborne Division. In the course of several field exercises, Westmoreland noticed a lack of experience on the part of small unit leaders when they took charge of their units when separated from their larger organization. Westmoreland was well aware of the Ranger School and had sent as many of his soldiers there as he could. Westmoreland realized though that not everyone could be sent to Ranger School for training. He therefore saw the need to improve small unit leadership within his division and sought a means to train soldiers in those positions. Members of his staff recommended to him that he could use those paratroopers of his division who were Ranger graduates to set up a training course on post to pass on their training to the small unit leaders.

Westmoreland bought into that idea immediately and established a training course. He chose Major Lewis Millett, a Ranger graduate and Medal of Honor recipient, to lead the training program. The training emphasized patrol techniques, intelligence gathering, improvised explosives, survival, land navigation, rappelling, use of friendly and enemy weapons, and hand-to-hand fighting. Following the classroom training, there were various field exercises, which involved a parachute insertion followed by the use of patrol, ambush, escape and evasion, survival, and sabotage techniques. As was true with the training at the ASTC and Ranger School, leadership responsibilities and team composition on the exercises rotated so that everyone got an opportunity to perform the various duties required and to work with as many other soldiers as conditions allowed. As the training progressed, a phase was added that involved the students being captured and enduring a period as a POW, with emphasis on resisting interrogation. The training was initially two weeks long but eventually was reworked to last four weeks. When choosing a name for the course, Recondo was developed as a combination of Reconnaissance and Commando, to stress the purpose and roots of the training.

Graduates of the training course were awarded an insignia that was in the form of an arrowhead pointing down (to signify an airborne insertion); the choice of the arrowhead was to be an indicator of the hunting and tracking skills of an American Indian. The insignia was white and black in color to indicate both daytime and nighttime operations. When worn on the fatigue uniform, the insignia was black and olive drab. Later, the work "RECONDO" was added to the upper, wide portion of the arrowhead. This insignia was designed to be worn on the breast pocket of a soldier's uniform.

Eventually, over time, Recondo training was adopted by other units in the army, including XVIII Airborne Corps, 82nd Airborne Division, the 25th Infantry Division, and others. In 1960, when Westmoreland became the Commandant of the U.S. Military Academy at West Point, he instituted a Recondo training course for cadets on a voluntary basis.

Between 1957 and November 1961, the U.S. Army activated nine Long Range Reconnaissance Patrol companies in the Active Army, virtually all of them in Germany in the various commands, including V Corps, VII Corps, Seventh Army, and the 3rd Infantry Division; one of these companies was in the Southern European Task Force in Italy. In 1965, two of these companies (one assigned to V Corps and one to VII Corps in Germany) were converted to Long Range Patrol units. Then in December 1967, eleven LRRP companies were activated in the various commands in Vietnam.

In 1966 Westmoreland, now the commander of American forces in Vietnam, ordered a Recondo training school be activated in Nha Trang. The cadre were members of the 5th Special Forces Group. The students consisted of American soldiers who volunteered and some Allied soldiers. Many of the Americans were assigned to LRRP/LRP units in their parent organizations. Some units that provided volunteers conducted pre-course training to prepare them for Recondo School.

The MACV course lasted for three weeks (about 260 hours of classroom and field instruction). The first part of the training concentrated on lots of physical training, patrolling techniques, first aid, land navigation, radio procedures, and weapons familiarity. The final exercise was an actual combat patrol led by the instructors. The first week of training was conducted on the school compound. The focus on this phase was physical training and classroom lectures dealing with map reading, intelligence reporting, and related subjects. The second week was spent outside the compound on Hon Tre Island in the South China Sea. The focus for this phase was familiarity with weapons, especially those used by the Viet Cong (VC) and the North Vietnamese Army (NVA), rappelling techniques from the ground and from a helicopter, ambushes, escape and evasion techniques, and other field activities. The focus of the third week was an actual combat patrol in the mountainous jungle between the massive naval air bases at Nha Trang and Cam Rahn Bay. As with the training for the Alamo Scouts and at Ranger School, students switched positions during the patrol to demonstrate what they had learned in the classroom phase. Graduates of this school were awarded the MACV Recondo patch, which was similar to the one designed when Westmoreland was at Fort Campbell, with the addition of a large "V" in the point of the arrowhead, to indicate the training had been conducted in Vietnam.

At the end of training, those soldiers who graduated gave written evaluations of the program, in an effort to improve any of the training. Some time after they returned to their respective units, graduates were re-contacted and asked to discuss how much of what they were taught was of the most use to them when they conducted actual operations and which was of the least use. This was done so that if any adjustments in the curriculum were needed, they would be based on comments from the field rather than by the instructors. In 1968, when Westmoreland was replaced by General Creighton Abrams, the MACV Recondo School was closed. During its operation, more than 2,700 American and 333 Allied soldiers had graduated from the school.

A careful comparison between the curriculums of the ASTC and the MACV Recondo School shows a marked similarity in what was taught. And so, even though the ASTC said it "just made up" its training curriculum, it apparently had it pretty much correct right from the beginning.

In 1968, the army was studying how to pull together all of the independent Long Range Reconnaissance Patrol companies into the Combat Arms Regimental System. The LRRPs did not have any single lineage or history. It seemed logical to designate them as Ranger units, but the answer was not as simple as that. The various Special Forces Groups of 1st Special Forces had all drawn their lineage and honors from the World War II Ranger battalions, so these were not available for use. The Korean War Ranger/Airborne units had been company strength and therefore were not color-bearing units. They could be used for honors but not lineage. The army ultimately decided to use 75th Infantry Regiment as the parent unit for the newly designated Ranger companies, which was done on 1 January 1969.

The 75th Infantry Regiment had been activated as the 75th Regimental Combat Team in November 1954 on Okinawa. It had derived its lineage and honors from the 475th Infantry Regiment, which had resulted from the deactivation of the 5307th Composite Unit (Provisional), more popularly known as "Merrill's Marauders," in August 1944. The 475th was deactivated in July 1945.

Fifteen separate Ranger companies, 13 of them serving in Vietnam, were part of the 1969 reorganization. The 75th Infantry was again deactivated more than three years later, on 15 August 1972.

Following the end of the Vietnam War, several divisions and one corps instituted their own Recondo Schools in order to train as many of their small unit leaders as possible in the techniques of long range reconnaissance patrolling and the other techniques that had been taught at the previously active Recondo Schools. Some of these schools were active into the 1980s. Several Reserve Officer Training Corps (ROTC) organizations on American university campuses also instituted Recondo training for their cadets. The various divisions and the corps issued their own versions of the Recondo insignia.

The Ranger units did not die, however. In 1974, the 75th Infantry was again activated as the 75th Ranger Regiment, first on 8 February with the 1st Ranger Battalion at Fort Gordon, Georgia, and then on 1 October with the 2nd Ranger Battalion at Fort Lewis, Washington. In 1984, following the invasion of Grenada, where the 1st and 2nd Ranger Battalions did in fact "Lead The Way," the 3rd Ranger Battalion and 75th Ranger Regiment Headquarters were activated at Fort Benning, Georgia. Finally, in mid-1986, the lineage and honors of the six Ranger Battalions from World War II were transferred from 1st Special Forces and the honors of the Ranger/Airborne companies from the Korean War were added with the lineage and honors of the 5307th and succeeding units, and presented to 75th Ranger Regiment.

Sources

Books
Baker, Alan D. *Merrill's Marauders*. New York: Ballantine Books, 1972.
Black, Robert W. *Rangers in World War II*. New York: Ivy Books, 1992.

Black, Robert W. *Rangers in Korea.* New York: Ivy Books, 1989.
Dilley, Michael F. *GALAHAD: A History of the 5307th Composite Unit (Provisional).* Bennington, VT: Merriam Press, 1996.
Gebhardt, James F. *Eyes Behind the Lines: US Army Long-Range Reconnaissance and Surveillance Units.* Fort Leavenworth, Kansas: Combat Studies Institute Press (Global War on Terrorism Occasional Paper 10), no date.
Gray, David R. "Black and Gold Warriors: U.S. Army Rangers During the Korean War." PhD dissertation, Ohio State University, 1992.
Hogan, Jr., David W. "The Evolution of the Concept of the U.S. Army's Rangers, 1942–1983." PhD dissertation, Duke University, 1986.
Ind, Allison. *Allied Intelligence Bureau: Our Secret Weapon in the War Against Japan.* New York: Curtis Books, 1958.
Ladd, James. *Commandos and Rangers of World War II.* London: Macdonald and Jane's, 1978.
Lanning, Michael L. *Inside the LRRPs: Rangers in Vietnam.* New York: Ivy Books, 1988.
Padden, Ian. *U.S. Rangers.* New York: Bantam Books, 1985.
Pugh, Harry. *Rangers: United States Army, 1756 to 1974.* Unpublished manuscript in author's collection.
Rottman, Gordon L. *US Army Rangers & LRRP Units 1942–1987.* London: Osprey Publishing, 1987.
Stanton, Shelby L. *Vietnam Order of Battle.* Mechanicsburg, PA: Stackpole Books, 1981.
Stanton, Shelby L. *Rangers at War: Combat Recon in Vietnam.* New York: Orion Books, 1992.
Zedric, Lance Q. *Silent Warriors of World War II: The Alamo Scouts Behind Japanese Lines.* Ventura, CA: Pathfinder Publishing, 1995.
Zedric, Lance Q., and Michael F. Dilley. *Elite Warriors: 300 Years of America's Best Fighting Troops.* Ventura, CA: Pathfinder Publishing, 1996.

Articles

Author unknown. "Ranger Units' Lineage, Honors Go To 75th Ranger Regiment." *Army,* June 1986.
Dilley, Michael F. "Training the Alamo Scouts." *Behind the Lines,* March–April 1995.
Dilley, Michael F. "The Alamo Scouts: World War II's LRRPS." *Patrolling,* June 1997.
Hughes, Les. "The Alamo Scouts." *Trading Post,* April–June 1986.
Jorgenson, Kregg P. "Long Range Recon Patrol: 1901." *Behind the Lines,* November–December 1992.
Ugino, Richard. "Historical Outline: Ranger Units." Unpublished paper in author's collection.
Wells, Jr., Billy E. "The Alamo Scouts: Lessons for LRSUs." *Infantry,* May–June 1989.
Zedric, Lance Q. "Prelude to Victory: The Alamo Scouts." *Army,* July 1994.

CHAPTER 10

The Son Tay Raid

In 1970, the nine-year Vietnam War was at a turning point. Two years earlier, after U.S. military and intelligence organizations had assured President Lyndon B. Johnson that the enemy forces were weak and losing the war, the communists launched a series of attacks known as the "Tet Offensive" that took everyone by surprise. Although the communist forces were soundly beaten on the battlefield, they had achieved a major propaganda and media victory. Later that year, Richard M. Nixon was elected president, largely by promising to end the war. After almost 18 months of secret and not-so-secret talks to fulfill that promise, Nixon authorized U.S. military forces to conduct a limited incursion into Cambodia in an operation aimed at the communist headquarters in the south, COSVN (the Central Office for South Vietnam). Public opinion, already hardened in rage against the government, exploded in rage at this expansion of the war. Even though the physical location of COSVN was overrun, no major victory resulted from the Cambodian incursion. More emphasis was then placed on "Vietnamization"—turning the prosecution of the war over to the South Vietnamese themselves in order to permit the withdrawal of U.S. forces. Negotiations between the U.S. and North Vietnam continued with both sides emphasizing, in different ways, the POW issue.

In and around the city of Hanoi, the capital of North Vietnam, were several holding stations and compounds that housed American pilots and air crew members as prisoners. These Americans had been shot down by anti-aircraft batteries or surface-to-air missiles (SAMs). As the

Americans were brought into the prison system, most of the "new guys" were taken to the isolated part of the old French prison, Hoa Lo, which they eventually called Heartbreak Hotel. Here they went through some initial interrogation and, when virtually all of them refused to provide more than their name, rank, service number, and date of birth, they were introduced to "the ropes."

Despite having signed the 1957 Geneva Accords on Treatment of Prisoners of War (a fact it denied throughout the war), North Vietnam made it a matter of government policy to brutalize and mistreat the Americans and, later, other Allied prisoners of war. This brutality was inflicted for even the slightest excuse. Initially, prisoners who refused to talk were accused of not having a proper attitude. Their wrists, forearms, and elbows were tied together behind their backs in an excruciating manner, and they were hoisted into the air by another rope tied around their wrists and thrown over a ceiling beam or through a hook; they were usually left in this position for hours—to consider their attitudes. Most prisoners were kept in solitary confinement in cramped, bare cells with no lights, toilet facilities, heating, or blankets. They were confined by painful leg stocks at night.

The prisoners' diet consisted of rice gruel or pumpkin sour, twice a day. This diet caused serious weight loss which led to weakness and low morale. Communication with other prisoners was forbidden and heavily punished when detected. Nonetheless, the prisoners did manage to communicate by tapping out Morse code messages on the cell walls. Later, a "tap code," based on a five-by-five matrix of the alphabet (minus the letter "K")—first developed during the American Civil War—was used, as it was easier to remember and send. New prisoners were quickly (and secretly) taught the code and how to "get on the walls" to communicate with their fellow inmates. This communication saved the lives and sanity of many despondent prisoners.

In addition to brutalizing the captive Americans, the government of North Vietnam steadfastly refused to acknowledge how many prisoners it held or who the prisoners were. Several public "shows" were staged for carefully chosen foreign press representatives or American individuals and groups opposed to the war, including Jane Fonda. At these shows, apparently well-fed, well-clothed, and well-treated prisoners were brought in to meet the foreign visitors.

Some of these shows backfired on the North Vietnamese. For example, Captain Jeremiah A. Denton, Jr., a U.S. Navy pilot, made a statement that he knew would be filmed; as he spoke, he blinked his eyes in Morse code to spell out the word "TORTURE" over and over again. (This was later detected by a sharp-eyed navy intelligence analyst.) Lieutenant Commander Richard A. Stratton, another navy pilot, marched about the room stiffly, staring straight ahead all the time and bowing very deeply—all to convey the message that he had been forced to be present at the "show." Many other prisoners were blunt (if sometimes secretive) when they could be, following the lead of Seaman Douglas B. Hegdahl. Whenever he was introduced to American war opponents, Hegdahl gave them the finger. Additionally, if he was put anywhere near food in the rooms, Hegdahl began to stuff his mouth with as much as he could get in before his captors forcibly removed him.

Later, the prisoners paid for many of their misbehaviors; but they were willing to take the punishment because they knew they had embarrassed the North Vietnamese and got the word out that they were not cooperating with them, despite whatever might be said by the communist government.

The presence of American POWs was a major war issue and continued to be so even after the peace talks began in Paris during the Nixon Administration. America adamantly demanded information about and access to its personnel held in prisons. The government of North Vietnam just as stubbornly refused to cooperate on the subject.

Based on various sources of intelligence (human sources as well as overhead photography from drones, planes, and satellites) and on information given by nine prisoners who were released between February 1968 and August 1969, many of the prison camp locations were known. Hegdahl, whose captors thought he was retarded, was ordered by the senior ranking officer of his camp to accept parole if it were offered to him. This order contradicted the Code of Conduct for Prisoners of War but was done deliberately because the Americans knew something the North Vietnamese didn't: Hegdahl had a photographic memory and remembered the personal data, shoot-down or capture dates, physical conditions, and locations of over 250 American POWs.

Eventually, Hegdahl was released in the last group of three in August 1969. The information he brought out with him was a veritable gold mine of intelligence.

Camp locations, including the names given by the POWs—names such as the Briar Patch, Rockpile, Skidrow, Alcatraz, the Zoo, the Hanoi Hilton (a reference to Hoa Lo) and others—were precisely plotted and photographed from overhead platforms. Data similar to that kept in intelligence target folders was collected, collated, and analyzed on each camp. Because of the way these prisoners had been released, the U.S. Government was reluctant to say publicly what it had learned from them about the horrible conditions of those who remained behind. The fear was that such public statements would lead to harsher treatment and a suspension of further prisoners released. Information about American POWs, however, became a very high priority for collection.

In early 1970, intelligence analysts of the U.S. Air Force's (USAF) 1127th Field Activities Group, stationed at Fort Belvoir, Virginia, began developing information about two camps west of Hanoi. One was located about 30 miles west, in a place called Ap Lo. The other was not as far west, only about 23 miles; it was located inside a walled area near the provincial capital of Son Tay. In late April, reconnaissance photographs showed increased outdoor activity at Son Tay. Imagery analysts concluded that the compound was being expanded.

Then, on 9 May 1970, a technical sergeant discovered something very unusual on a photograph of the Son Tay camp: he thought he read a message from the pattern of clothes and other objects that spelled out a message based on the "tap code." The message seemed to suggest that six of the prisoners were planning to escape and wanted to be picked up in the foothills near a village eight miles southwest of the camp. The sergeant had, prior to this Saturday morning, assembled fairly conclusive evidence of the presence of 55 POWs at Son Tay. Now it appeared that several of them were planning to go over the wall and were asking for help.

During the next two weeks, briefings and meetings were held in various USAF organizations to discuss the sergeant's analysis of the

Son Tay imagery. Not everyone agreed with his analysis, although the supervisory staff of the 1127th did and they continued to push for a mission to help rescue the escaping POWs. Following a briefing to the USAF Deputy Chief of Staff Plans and Operations (DCSP&O), a meeting was arranged for 25 May so that the analysts of the 1127th could brief an army brigadier general named Donald D. Blackburn who had the interesting, but obscure, title of Special Assistant to the Chairman (of the Joint Chiefs of Staff, JCS) for Counterinsurgency and Special Activities. Most people referred to Blackburn as the SACSA.

General Blackburn came to the position with many years of experience in unconventional operations. In 1942, as a young army officer, he was stationed in the Philippines. When the Japanese overran Luzon, Blackburn was one of several officers who disobeyed the order to surrender. Instead, he took to the hills in the north of Luzon and established a guerrilla army among the Igorot headhunters. In the late 1950s, he transferred into the U.S. Army's Special Forces; he organized and sent to Laos the "White Star" teams that trained the nascent Laotian Army. In 1965, he was selected to head the Studies and Observation Group (a cover name for the Special Operations Group, or SOG), which directed much of the unconventional war in Vietnam, particularly controlling an all-service force that conducted many of its operations out-of-country. Blackburn was later an assistant division commander of the 82nd Airborne Division. In army circles he was a vocal supporter of unconventional operations.

Blackburn and his assistant, an army colonel named Edward E. Mayer, Chief of SACSA's Special Operations Division, listened intently to the briefing. By now the 1127th had included in its briefing some details for how a rescue mission could be executed. These details included putting an army Special Forces team within striking distance of the camp and sending a human agent into the Son Tay area to assess the situation. SACSA had control over assets that could, if approved, arrange for all aspects of the proposed plan. Blackburn liked the idea for several reasons. He had other plans for special operations in North Vietnam and a success like this could boost his ability to get these other plans approved. The most important reason why he liked this plan, however, was that if it could be pulled off, the rescued prisoners could speak out about conditions in the camps, unhindered by any concern for violating the "conditions of their release."

As the meeting progressed, Blackburn suggested that if they could get six prisoners out, maybe they could hit Son Tay and Ap Lo at the same time and get all the prisoners at each camp out. What a coup that would be! However, Blackburn didn't have the authority to approve such a mission; that would have to go to the Chairman of the JCS. As the meeting ended, Blackburn promised to get back to the 1127th after describing this brief to his boss. Soon after the USAF intelligence analysts left, Blackburn and Mayer briefed General Earle Wheeler, the Chairman of the JCS. Wheeler's immediate reaction was, "Jesus Christ, Don. How many battalions is this going to take?" Nevertheless, he gave Blackburn a go-ahead to put a plan together and be prepared to brief a meeting of all the chiefs on 5 June.

Over the course of the next six weeks, Blackburn and Mayer were busy briefing the Joint Chiefs, receiving initial approval, and forming a feasibility study group to look at supporting intelligence and possible operational options. The raid was given the code name *Polar Circle.* One of Blackburn and Mayer's most important tasks during this period was to find the correct people to inform about the operation, who could support them, yet still keep the list of those who did know down to the absolute minimum. Blackburn was convinced that he would only get one chance at an operation like this and any breach of security would be disastrous.

A decision that was made early in the planning phase was to fly the raiding force to Son Tay in helicopters from somewhere in Thailand, probably U Dorn Royal Thai Air Base. Such a flight would require the Air Group to "thread the needle" of all the intervening anti-aircraft radar sites in order to get in undetected. The National Security Agency provided detailed information about North Vietnamese radar networks, with blind spots identified. Precise routes had to be plotted, the Air Group had to fly as close to the ground as possible, and an in-flight refueling would be required. It was also about this time that an idea was raised to conduct a diversionary air raid over or near Hai Phong, on North Vietnam's east coast, to draw all attention to it and away from Son Tay.

On 10 July, Blackburn and a member of his feasibility study group briefed the new chairman, Admiral Thomas Moorer, and the other chiefs, seeking approval to begin assembling a force to carry out the raid. At this, Moorer's first meeting as chairman, several new issues were raised. Analysts were certain that the camp at Ap Lo was empty; they had no idea where the prisoners had been sent but they knew conclusively that they were gone. At Son Tay, 61 prisoners had been positively identified by name as being in the compound. Additionally, a compound south of Son Tay that looked very similar to the prison camp was identified as a "secondary school." The JCS gave Blackburn approval to recruit a raiding force and begin training.

That following Monday, 13 July, Blackburn and Mayer went to Fort Bragg, North Carolina. They had lunch with Colonel Arthur D. Simons, the Assistant Chief of Staff G-4 (Logistics), XVIII Airborne Corps, and asked him if he was interested in leading a "very sensitive mission," without giving him any further details. Simons' answer was "Hell, yes, let's go. I don't need to know any more about it."

Blackburn was not surprised at the answer from the special forces legend, whose nickname was "Bull." Simons' unconventional career stretched back to World War II, when he was an original member of the 6th Ranger Battalion in the Southwest Pacific, commanding B Company. His operations with the Rangers formed the start of his legend, as he led several operations behind Japanese lines to conduct reconnaissance or raids at the direction of Sixth Army Headquarters. He transferred to Special Forces in the late 1950s and eventually was chosen by Blackburn to lead the "White Star" teams in Laos. He later commanded Special Forces units in South America and Vietnam. Most of the men who served with him were willing to go anywhere he led them.

Blackburn, Mayer, and Simons then discussed who else should be on the operation. At least two other names were added to the team: Lieutenant Colonel Elliot P. Syndor and Captain Richard J. Meadow, both currently assigned to the Ranger Department, U.S. Army Infantry School, Fort Benning. Meadows had served with Blackburn and Simons before in Special Forces assignments. He was a sergeant on a SOG team when Blackburn was chief of SOG. He led a SOG operation into Laos and captured a North Vietnamese artillery piece. For this incredible mission he received a battlefield commission, the first awarded during the Vietnam War. He later served with Simons in Panama and impressed him with his unconventional skills and his nerve. Syndor, too, had an extensive background in Special Forces and had served with Simons before. Simons considered him one of the best Special Forces soldiers he knew.

Before leaving Fort Bragg, Blackburn and Mayer determined that training areas for the raiding force were not available there and decided to use Eglin Air Force Base, Florida, the home of the Air Force Special Operations Force. Later that same day, Blackburn and Mayer received approval from Admiral Moorer to use Eglin. A message was sent to the USAF at Eglin to name a mission commander for what was then called Operation *Ivory Coast*. The ad hoc organization that would conduct this operation was given the innocuous name "Joint Contingency Task Group" (JCTG). Before long, USAF Brigadier General Leroy J. Manor was named to command the JCTG.

Manor had over 350 combat air missions, flying in World War II and Vietnam. He was an interesting compliment to Simons. He was quiet yet had the same degree of efficiency as the Special Forces colonel. His recent experience in Vietnam had brought him into contact with many of the best fliers in the USAF, both fixed-wing and helicopter. He had trained many of them to support SOG's unconventional operations.

Unknown to all the planners, on 14 July all American prisoners were removed from the camp at Son Tay. The camp's well had run dry and recent flooding from the nearby Song Con River threatened the compound with water that came to within two feet of the prison camp's wall.

In Washington on the 14th, Blackburn told Manor and Simons everything he knew about the planned raid on Son Tay. Manor would be the overall JCTG commander and Simons would be his deputy. Additionally, Simons would train and lead the raiders (army and air force), while Blackburn and Mayer would run interference in Washington. That same day Manor was given a "To whom it may concern" letter from the Chief of Staff of the USAF, ordering all air force commanders to

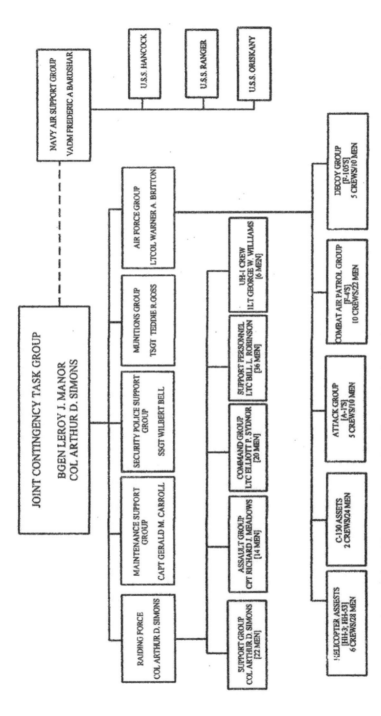

Organization chart of the ad hoc Son Tay raiding force. The dotted line indicates a coordination relationship rather than a direct command line. Because the raiders were kept in the dark and not told the real mission until just before launch time, they referred to themselves as "mushrooms". (Source unknown)

give Manor whatever he needed, "no questions asked." Field training for the army raiders was to begin on 9 September. The first window of opportunity for the raid, based on weather and light data, was between 20 and 25 October, since North Vietnam would be in its monsoon or rainy season until just before then. Finally, Manor and Simons were told to have their operational plan ready by the end of August.

Later that summer, after bringing Syndor and Meadows on board, Simons addressed a theater crowded with Special Forces troopers at Fort Bragg. He couldn't tell them much except that the mission would be hazardous and there would be no extra pay for temporary duty; more than half of the initial group left. In addition to those who came to the theater, sergeant majors of several Special Forces units on post forwarded the names of combat veterans in their units who might be good candidates for a hazardous mission.

During the next three days, Simons and the raiding team doctor (an army Special Forces-qualified lieutenant colonel named Joseph Cataldo) and several others screened and interviewed the remaining volunteers. Medical and personnel records were reviewed. The list of questions in the interview would tell those being interviewed nothing about the operation. The questions covered a wide variety of topics that were more designed to give Simons and Cataldo a chance to listen to the volunteers, assess their potential physical abilities (such as being able to carry a prisoner from the compound to a helicopter), and observe them answer questions that embarrassed them and even put them under stress. Eventually, they got their final list down to 15 officers and 82 enlisted men.

Most of these men were combat veterans, although six of them were not. Typical of the kind of soldier selected was Master Sergeant Galen C. Kittleson. During World War II Kittleson had, as a 19-year-old PFC, served on the Nellist Team with the Alamo Scouts, the special reconnaissance unit created by Lieutenant General Walter Krueger, commander of Sixth Army. Kittleson was on the two operations of the Alamo Scouts that liberated POW camps, the first in October 1944 at Oransbari, New Guinea, and the second in January 1945 at Cabanatuan on Luzon. He left the army after the war but came back in during the mid-1950s and later joined Special Forces. In March 1967, while in Vietnam, he was on an unsuccessful raid to free James N. ("Nick") Rowe, a U.S. Special Forces officer held prisoner in the Delta region of South Vietnam. In 1978, when he retired from the army as a sergeant major, Kittleson was the last of the Alamo Scouts on active duty.

Then Simons sent representatives to Eglin Air Force Base to pick a training area. As coincidence would have it, the area that was finally used was near Auxiliary Field 3 (known locally as "Aux 3" and "Duke Field"), the same area that was used to train the pilots for the Doolittle Raid in 1942. Helicopter pilots and crews were picked next, from the cream of Air Force Special Operations. Manor led this effort because of his position and because he knew who most of the good Special Operations pilots and crews were. Like Simons, he knew who he wanted to take with him on an operation like this. Lieutenant Colonel Warner A. Britton was the senior helicopter pilot and his main job was to train the air crews so they would be ready when joint training began with the Special Forces troopers, in late September.

While the training area at Eglin was being established, Blackburn convened a planning group in Washington for a solid week of making decisions and reviewing current intelligence on the target area. During this period, the Defense Intelligence Agency (DIA) briefed the group that Son Tay had not seemed very active since early June. The operational planners devised long flight missions to train the helicopter crews, while the ground force was to concentrate on physical training, team and patrol drills, and marksmanship—a lot of marksmanship. Two elements, to perform logistics and armorer duties for the entire force, were selected and given broad mission orders to begin drawing equipment—or buying it or stealing it, if necessary.

Both groups, air force and army special forces, brought a broad background to the force, but Simons intended to focus their effort. The training and drills he put them through were all based on what they would do in the actual operation, plus some contingency training, survival, escape and evasion, and a healthy dose of emergency medical treatment. So far, only four people in the ground force knew what the real target was: Simons, Syndor, Meadows, and Cataldo.

A cover story was developed to give the men sufficient information about the operations but not be too close to the truth. They thought they would be raiding a village to rescue diplomats and other embassy personnel held hostage. The operational name, *Ivory Coast*, led them to think they were headed for Africa, and Simons let them think whatever they wanted. As time passed, many of the men figured out that they would be hitting a POW camp, probably somewhere in Southeast Asia, although none imagined it would be in North Vietnam.

Soon they began training with a collapsible mock-up of the village, made of wood and target cloth. It could be, and was, easily assembled in the dark (when satellites couldn't photograph what they were doing) and be disassembled and stored away before first light. They also met "Barbara," a detailed terrain model mock-up of the prison compound and the surrounding area with special features to simulate various degrees of available light. It was built for them by the Central Intelligence Agency.

As the training progressed, Simons examined the raiders' actions more closely. He began to look for ways to take out the tower guards more efficiently and get part of his force into the compound quicker. The first part of the plan called for a helicopter to fire at the guard towers with M-60 machine guns, taking out the guards. The other part was difficult because of the space inside the compound; Simons thought that landing a helicopter inside would give the prisoners less time to be exposed to fire by their captors before his raiders could get to them and would also give his force the added advantage of surprise and shock.

At first an army UH-1—"Huey" or "slick"—helicopter was used with a select army crew. This aircraft was dropped because it did not have the carrying capacity needed, could not be refueled in-flight, and did not have the range to fly from Thailand to Son Tay. The solution to these problems was to use the HH-53 (Super Jolly Green Giant) for transport since it fulfilled all the criteria the Huey did not. The air crews then suggested crashing an HH-3 (Jolly Green Giant) helicopter instead of the Huey. Although smaller than the HH-53, it was just as rugged and durable, and its blade length would give them just enough clearance. This also meant that the HH-3 that was to be crashed would have to be destroyed, before the raiders pulled out.

Simons eventually divided his ground force into four teams, headed by himself, Meadows, Syndor, and Lieutenant Bill L. Robinson. Meadows'

Part of the Son Tay raiding force. (Library of Congress)

team (code-named "Blueboy"), the smallest, had 14 men. Their job was to be on board the helicopter that would crash inside the compound. Simons' team (code-named "Greenleaf"), of 20 men, would be next, securing a perimeter for Syndor's team. This team (code-named "Redwine"), of 22, was the largest to land and would breach the prison outer wall with a satchel charge. Robinson led the largest team overall, 36 men. They acted as a back-up pool and had to know what virtually every man on all the other teams did, in case any of them could not make the final operation. Some shifting did occur, although not much. For example, as late as just a few days before the raid, one of the men on one of the first three teams had to be replaced. Robinson's team also supported the other three in training in various ways, such as setting up and tearing down the simulated village and acting as the enemy force.

Obtaining the equipment and ammunition that the air and ground groups needed for their extensive training and eventual mission was a monumental task. There were items that were in scarce supply in the military system; some had to be purchased commercially because that was the only source. Reliable ammunition was a persistent problem and not just because of the variety of weapons that would be used. So many exceptional purchases or requisitions were made that Simons was concerned that his logistics "tail" would compromise the mission.

Once most of the logistics problems had been resolved, live fire rehearsals began, both day and night iterations. On 28 September, the air and ground groups began joint training. The tempo of three assaults in both day and night hours continued. When not rehearsing, the teams walked through their actions in and around the compound so that each man knew exactly where every other man was and would be. "Barbara" was a constant training aid in these sessions. By now the Special Forces raiders had fired so many rounds in so many rehearsals (between 150 and 170) that Simons was certain they knew where each round would land.

On about 6 October, Manor and Simons decided their men were ready for the first real-time "full-scale, live-fire, night dress rehearsal." The air portion of this rehearsal may have been the most difficult, as would be the actual air execution on the raid, because the flight involved many out-of-the-ordinary hazards, just in reaching the target. The accompanying and re-fueling C-130s had to fly at 105 miles per hour, just over stall speed, and with 70 per-cent flaps so that the HH-3, the slowest helicopters, which had a top speed of about 105 miles per hour, could keep up. The A-1s and other planes that would also be flying this leg couldn't fly that slowly and so the pilots had to execute S-turns and circling maneuvers to stay with the remaining aircraft. This would have to be done at low level, all the while evading radar and in complete radio silence.

In Washington, Blackburn continued the round of briefings to receive approval for the final operational launch. Political meetings abroad by President Richard M. Nixon and his National Security Advisor, Henry A. Kissinger, continued to center around obtaining more information about or getting the release of American prisoners. While he was in Europe on a trip, President Nixon was first briefed on the Son Tay mission and he authorized the planning to proceed.

In early October, the raid was postponed for a month because of political considerations. This extra time did little to improve intelligence about Son Tay. Many satellite and SR-71 photographs were still useless because of heavy cloud cover. Drones were sent sparingly for fear that too many would signal a particular interest. DIA analysts reported that there appeared to be less activity at Son Tay than they had seen in the past and they did not observe recent outdoor activity by the POWs. They concluded that perhaps the prisoners were being punished.

On 1 November, Blackburn, Manor, and Simons left for the Pacific area on a final round of pre-raid coordination. This included visits to Hawaii and Saigon, capital of South Vietnam, and a separate trip to brief Vice Admiral Frederic A. Bardshar to arrange for the U.S. Navy carriers on Yankee Station in the Ton Kin Gulf to launch the diversionary raid on Hai Phong Harbor. While they were gone, analysis of new photographs indicated an increase in activity at both Son Tay and the "secondary school" to the south.

On 12 November, the White House gave a conditional authorization for the raid to go ahead on the evening of 21 November; a final authorization briefing for the president was scheduled for 18 November. Also on the 12th, the first C-130 Combat Talon aircraft departed, bound for Ta Khli Royal Thai Air Base, the designated "isolation area" for the final briefings.

Blackburn and Mayer were still putting out fires in Washington. The latest involved asking the USAF to ground all its HH-53s for a "potentially catastrophic technical problem" until further notice. This was caused by CIA requests to use aircraft earmarked for the raid. General Creighton Abrams, Military Assistance Command Vietnam commander, who knew about the raid, declined to release the helicopters to the CIA but couldn't tell them why; local CIA officers were trying to find ways to bypass Abrams and get the helicopters. On 17 November, the raiding force arrived at Ta Khli.

Just before noon on 18 November, Admiral Moorer briefed President Nixon on the final details of what was now called Operation *Kingpin*. The president approved the raid and a message to "execute" was sent to Manor in Thailand. At about the same time, Typhoon Patsy slammed into the Philippines with winds of 105 miles per hour, and turned west toward North Vietnam.

At 1630 the next afternoon (0400 on the morning on 20 November in Thailand), the Director of the DIA told Blackburn that there might not be any prisoners in Son Tay, that the compound looked empty. A meeting was quickly arranged with Admiral Moorer. A decision was made to meet at breakfast early the next morning with the Secretary of Defense, Melvin Laird, and brief him. During the night, at just about 0400, Washington time, Manor notified the Pentagon that he had moved the raid up by 24 hours because of Typhoon Patsy, which was now predicted to hit North Vietnam on the night of the 21st. The raid would launch at 1032, Washington time, unless he received orders to the contrary.

What Manor did not say in his message was that he had increased his air group. He decided to add five F-105s to escort the raiding force into North Vietnam to act as decoys once they were over the target—he hoped that they would draw the attention of anti-aircraft batteries and SAM crews. After he sent this message, Manor departed Thailand for Monkey Mountain, south of Da Nang in South Vietnam, a communications center where he could receive traffic from the raiders at Son Tay as well as the National Military Command Center at the Pentagon; he would remain at Monkey Mountain until the raiders left Son Tay.

Blackburn had spent much of the night reviewing recent intelligence reports with DIA analysts. Human sources not specifically targeted to collect data on POWs provided information indicating that there were no prisoners at Son Tay. Infrared imagery of the compound, however, indicated activity and photographs showed that the compound had recently been enlarged. Information provided by a special source showed that information would soon be released that at least eleven POWs had died since the beginning of the year, one as recently as early November; this gave added urgency to the mission. Blackburn finally told the analysts to make one comprehensive review of what they had and to give him an unequivocal answer in the morning. At a meeting before breakfast, the intelligence analysts told Moorer and Blackburn that they had conflicting information but recommended that the raid go. This was passed to Defense Secretary Laird at breakfast and he agreed. The raid was now a "go" for the final time.

In Thailand, the raiders—who still did not know the real mission—had been going through last-minute briefings, re-zeroing weapons, and holding shooting practice for two days. On the afternoon of 20 November, with the entire force assembled in a theater, Simons spoke briefly to the men. "We are going to rescue 70 American prisoners of war, maybe more, from a camp called Son Tay. This is something American prisoners have a right to expect from their fellow soldiers. The target is 23 miles west of Hanoi." Simons recalled later, "You could hear a pin drop. I want to tell you it got pretty quiet. Very quiet." Then, as one, the men stood and applauded. "You are to let nothing, nothing interfere with the operation," Simons went on. "Our mission is to rescue prisoners, not take prisoners."

Simons offered to let any man who did not want to go on the mission leave and nothing would ever be said about it. No one left. After a few more comments, Simons turned the briefing over to Syndor. As he left the room Simons heard one of the men say, "Jesus, I'd hate to have this thing come off and find out tomorrow I hadn't been there."

Soon after, the force loaded onto a C-130 and flew to U Dorn. On Yankee Station, as the raiders flew to their jumping-off point, navy pilots from three carriers, USS Oriskany, USS Ranger, and USS Hancock, were getting the word that they would be flying missions armed only with flares. Bardshar's orders were very specific: "No air-to-ground ordnance is authorized with the exception of flares carried by Strike aircraft, and the Rockeyes [cluster ammunition] and guns carried aboard the Rescue Combat Air Patrol." At the last minute, Bardshar did authorize a few aircraft to be armed with "radar-homing missiles to suppress ... air defense batteries."

Just before 2300 in the evening, the helicopters with Simons and his men on board took off from U Dorn. They were led by two C-130 Combat Talons. Simons told his men to wake him when they were 20 minutes out from Son Tay, then went to sleep. At about the same time a refueling C-130 left Ta Khli, and the A-1s, F-105s, and also F-4s left from Nakhon Phanon—all headed to Laos, flying without navigation and cabin lights and in radio silence.

Colonel Arthur D. "Bull" Simons briefs the raiders of the Son Tay operation prior to attack. (Library of Congress)

Thick clouds covered the route through Laos, making visual contact difficult; at each break in the clouds the air convoy resumed formation. Over central Laos refueling began. One of the helicopter pilots later described this operation: "Well, daytime is pretty easy. Nighttime with the lights on is not too bad but nighttime with no lights and no radio will damn near give you a heart attack." Refueling was accomplished without a hitch, the last helicopter beginning its refueling at H minus 55 minutes. At the same time, on Yankee Station, navy jets took off to begin the diversionary raid.

As the air convoy crossed into North Vietnam, the various aircraft continued their delicate airborne dance, twisting and turning through the blind spots in the air defense radars as they approached Son Tay from the west. The final checkpoints in the flight were approaching: the Black River ten miles from the compound, Finger Lake (the raid holding area, where aircraft would go after dropping their raiders, waiting to be called back to pick up them with the freed POWs) seven miles out, and the Song Con River where it turned north, just two miles out. The air convoy began to split apart, each aircraft heading off to perform its mission.

Just as they crossed the last checkpoint, at exactly 0217, the sky to the east over Hai Phong lit up. The U.S. Navy had arrived, although the raiders could not see that far. Right on cue, one minute later the Combat Talons near Son Tay began dropping flares, lighting up that area. The A-1s dropped combat simulators, to make any forces in the area think that a fire fight on the ground was in progress. The F-4s flew screaming missions for all the other aircraft while the F-105s hoped to attract the attention of SAM batteries away from the helicopters to themselves.

Major Herbert Kalen brought the HH-3 (carrying "Blueboy") toward the Son Tay compound, preparing to crash land inside. Kalen maneuvered the helicopter so that the gunners manning M-60s with all-tracer rounds could take out the guard towers on the wall but ignore the tower over the gate. This tower was on top of a metal shed. They were told in training that prisoners were put in this shed as punishment and Kalen's gunners (Kittleson on the port side and another raider on the starboard gun) were to take no chances that a prisoner was in the shed.

As he cleared the trees Kalen hit a clothesline. His careful maneuvering was gone, and it was all he could do to make sure he landed upright. Just before touchdown the blades cut unto a 10-inch-thick tree, twisting the helicopter sharply to the right before it slammed into the ground. The mattresses on the deck inside the helicopter, to absorb the blow of the hard landing for Meadows and his team, didn't help much. A fire extinguisher was knocked loose, hitting the flight engineer (Technical Sergeant Larry M. Wright) so hard that it broke his ankle. Meadows was to have been the first out of the helicopter but the force of the crash landing threw Lieutenant George W. Petrie out first. Meadows followed, leading his men. When he was clear of the wreck, he knelt, holding a bull horn, as his men headed for the cells and the front gate.

The next helicopter out from the compound, piloted by Lieutenant Colonel John Allison, with Syndor's team ("Redwine") on board, had watched the aircraft in front of them as it swung into action at 0218. Kalen's helicopter had started down, then came back up and pulled ahead. Allison realized that Kalen had made a mistake that they had been warned about in training—he had almost landed at the "secondary school," which looked a lot like Son Tay. There was even a nearby canal that could be mistaken for the Song Con River. Then Allison noticed that Britton's helicopter (with Simons' team, "Greenleaf," on board) veered off to the right. Britton landed at the "secondary school." Allison realized that they had drifted southward. He corrected his line of flight and headed for Son Tay.

When he arrived, Allison prepared to land just as Syndor passed the word to his men to execute Plan Green. This was one of several contingencies that Simons had planned, each presupposing that one of the helicopters with Special Forces on board did not make it. Allison told his gunners to take out the towers that should have been Britton's target and touched down. Syndor and his men spread out quickly. Master Sergeant Joseph W. Lupyak led the way to several buildings outside the compound that housed guards—those off-shift who were sleeping and those on-shift who were not on post but were still awake. Master Sergeant Herman Spencer moved to the compound wall and set a satchel charge. He was to blow a hole for them to enter the compound. Allison lifted his helicopter off just before the charge blew.

It did not take Simons long to realize they were in the wrong place. He could see barbed wire on the compound walls and did not expect to; he could not hear Meadows talking on the bull horn, as he had expected. As a fire fight broke out around him, Simons told his radio operator, Staff Sergeant Walter L. Miller, to recall the helicopter and tell Syndor to go to Plan Green. Then he told his signal specialist, Staff Sergeant David S. Nickerson, to turn on a strobe light to mark the landing zone for Britton. Finally, he turned his attention to the fire fight around him.

Meanwhile, inside Son Tay, Meadows was talking into his bull horn, saying, "We're Americans. Keep your heads down. Americans. This is a rescue. We're here to get you out. Keep your heads down. Get on the floor. We'll be in your cells in a minute."

The raiders on Meadows' team were running through the building, shooting at waist and shoulder level, figuring that the Vietnamese would not understand, as they hoped the POWs would, the instructions over the bull horn. Just as Meadows' radio operator called to tell Simons that they were inside the compound, a large explosion knocked Meadows down.

Syndor's team came pouring through the hole they had just blown. Syndor quickly told Meadows about Plan Green and the two team leaders immediately began to coordinate the actions of their men.

Simons' attention was caught by a moving light in the sky and a loud *wooooosh*. At first he thought a plane had been hit, but when the light kept moving, he decided it was a SAM, streaking skyward in search of a plane. Almost as suddenly there was a big explosion at the compound.

THE SON TAY RAID • 119

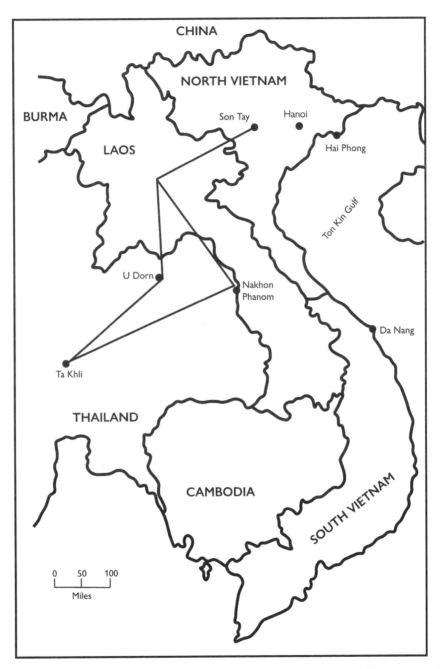

Air routes taken by the Son Tay raiding force. (Map by author, digitized by Casemate)

Simons knew his men did not have explosives and assumed that a stray round had probably hit some gasoline barrels. He and his men rushed into the compound (Simons was never sure how the hole in the wall they rushed through got there) and shot at everything that moved. For five minutes the fighting went on.

Back at Son Tay, Master Sergeant Thomas J. Kemmer and Petrie searched what they believed was an administration building. They ran from room to room. On the other side of the compound, Captain Udo Walther led a team through the cells, shooting several Vietnamese soldiers as they went. Outside the compound, Cataldo had established a medical screening station. In the center of the compound, Meadows was hearing the same message from his various search teams: "Negative items. Negative items." It was a message he had not expected to hear; it meant there were no prisoners.

Simons was also getting a surprise he definitely did not expect: the people his men were shooting at were tall, many of them close to six feet, some taller. Some were Asian but not all of them. They wore uniforms different from what he knew North Vietnamese soldiers wore. The fight was too hot and heavy for any of his men to determine with any success who they were fighting but afterward they speculated that they may have been Russian or Chinese.

Simons got a radio message that his "ride" was inbound, and he pulled his men back to the landing zone, fighting all the way. Britton landed under fire and stayed until he was sure everyone was on board. In the short trip to Son Tay, Simons radioed in an air strike on a road bridge between the "secondary school" and Son Tay ("insurance" he called it later), and called Syndor and Meadows to tell them he was on the way and to revert to the basic plan. In less than a minute from pickup, he and his men landed outside the compound at Son Tay. Britton had just completed three

Captain Richard J. Diamond, leader of the Assault Group of the Son Tay raid. (Library of Congress)

combat sorties in less than nine minutes! His navigation error was probably one of the most fortuitous "accidents" anyone on the raid had ever seen.

As Simons and Meadows linked up in the compound there were still several small fire fights going on. When Meadows told his colonel what he was hearing from his search teams, Simons was stunned. Meadows left to conduct a thorough search of his own. When he got back, the result was the same: "Negative items."

Without hesitation, Simons told his men to prepare to move to the landing zone for extraction. He set up security and sent one of his photographers back into the camp to take pictures of the empty cells. Then he told Miller to radio for an air strike on a bridge north of the compound and to bring two of the helicopters back from the holding area. Nickerson fired a flare to aid the pilots in finding the correct compound. Fourteen minutes had elapsed since the gunners hosed the guard towers.

When Britton landed outside the compound, Simons sent Meadows' team (without Meadows), Kalen and his crew, and part of Syndor's team (26 in all) on board. Britton lifted off and Simons radioed for the next helicopter. As the remaining 33 raiders waited for Allison to bring in his helicopter, Meadows went back to the compound one last time to affix demolition charges to the crashed HH-3. Then he dashed back out to wait for Allison.

In the sky overhead, SAMs continued to fly but they hit nothing. One came in the general direction of Britton's helicopter but his crew chief spotted it and yelled a warning into his radio. Britton banked sharply to avoid the missile. While they waited, the raiders on the ground spotted several trucks headed in their direction. They fired several LAW (Light Anti-tank Weapon) rockets at the truck, taking out this last threat to the raiders. When the helicopter landed, the raiding forces Marshaling Area Control Officer counted everyone aboard (as he had for the first helicopter) and passed the count to Simons. Allison pulled his helicopter off the ground 27 minutes into the raid. Six minutes later the HH-3 blew up inside the prison camp.

Everyone (Special Forces, air force crews, and navy pilots) made it back alive. Two raiders (Kalen's flight engineer and one Special Forces raider) were wounded and one F-105 was shot down after the raid, but one of the HH-53s pulled the crew out and brought them back. By the time they all arrived in Thailand, General Manor had returned from Monkey Mountain and he greeted them as they landed.

Several years later, details of a program known as Operation *Popeye* (and several other names) were revealed. *Popeye* involved attempts to alter weather patterns over North Vietnam. Specifically, it tried to increase the seasonal rainfall in northeastern Laos, hoping to cause heavy flooding in western North Vietnam. Most of the *Popeye* missions in 1970 occurred between March and November.

The Son Tay raid's After Action Report (written by Manor, who did not know about *Popeye*) said that five years' worth of typhoons hit the area of North and South Vietnam and Laos in the two months prior to the raid. From Blackburn down, no one involved in the raid planning was aware of *Popeye*. No one really knows for certain what influence *Popeye* had on the flooding that contributed to the decision by the North Vietnamese to move the prisoners out of Son Tay in July 1970.

It had been a near-perfect raid. Even the landing of Simons and his team in the wrong place was somehow right. The raiders had achieved complete surprise and, in Simons' words, the "attack was pressed with great violence—because surprise doesn't work if you don't use violence and speed." And yet, they came away empty-handed.

Had they failed? There is a school of thought that contends they did not fail, pointing to the precision of the raid itself and, more important, from their perspective, what happened to the POWs who remained in North Vietnam. Conversely, others believe the raid did fail, pointing to the failure of intelligence to give complete answers for the planners and raiders. Part of the intelligence failure, they say, was relying on merely technical means to collect information, and then limiting those means to mostly imagery. No human source was targeted to collect information about Son Tay and there was no apparent attempt to use signals intelligence except to map the radar blind spots. This school also blasts the more than six months that elapsed between the imagery with the "come and get us" message and the raid itself, while the planners tinkered with the plan to make it perfect. They even fault those who reviewed the plan, up to the president, for failing to ask tough questions and demanding tough answers. It seemed they had forgotten the words of General George S. Patton, Jr., who said, "A good plan violently executed now is better than a perfect plan next week."

What happened to the POWs as a result of the raid was that their captors realized their vulnerability. To compensate, they pulled all the prisoners from the outlying prison camps into one place: Hoa Lo prison, renamed Camp Unity by the POWs. Whereas before, many of these men had spent years in solitary confinement, now they were crowded 20 or more into a room. Now they could organize and act as a group; they could help one another, especially the sick and despondent; they were a military unit once more—the Fourth Allied POW Wing. It would still be more than two years before they were freed but the prisoners knew about the raid on Son Tay—and they were proud and glad that General Manor, Colonel Simons, and their men had tried to get them out.

Sources

Books

Bowers, Frank R., B. Hugh Tovar, and Richard Schultz, eds. *Special Operations in U.S. Strategy*. Washington, D.C.: National Defense University Press, 1984.

Brown, Ashley, and Jonathan Reed, ed. *The Unique Units* (A volume in The Elite series). London: Orbis Publishing, 1989.

David, Heather. *Operation: Rescue*. New York: Pinnacle Books, 1971.

Hubbell, John G, Andrew Jones, and Kenneth Y. Tomlinson. *P.O.W.: A Definitive History of The American Prisoner-of-War Experience in Vietnam, 1964–1973*. New York: Reader's Digest Press, 1976.

Kelly, Orr. *From A Dark Sky: The Story of U.S. Air Force Special Operations*. Novato, CA: Presidio Press, 1996.

Lipsman, Samuel, ed. *War in the Shadows* (A volume in The Vietnam Experience series). Boston: Boston Publishing, 1988.

Schemmer, Benjamin F. *The Raid*. New York: Harper and Row, 1976.

Vandenbrouke, Lucien S. *Perilous Options: Special Operations as an Instrument of U.S. Foreign Policy*. London: Oxford University Press, 1993.

Articles

Bowers, Everett P. "Expert POW Liberator." *Military History*, October 1996.

Conroy, Michael R. "POW Rescue Game Plan." *Vietnam*, unknown issue.

Dilley, Michael F. "Rescue at Cabanatuan." *Behind the Lines*, September–October 1993.

Hanley, Kipp. "Arctic Angels' Soaring Again." *Military Officer*, February 2023.
Hasenauer, Heike. "A Special Kind of Hero." *Soldiers*, November 1995.
MacDougald, N. W. "Who Dares Wins." *Soldier of Fortune*, June 1979.
Martin, David C. "Inside the Rescue Mission." *Newsweek*, 12 July 1982.
Schemmer, Benjamin F. "Requiem for a Warrior." *Soldier of Fortune,* November 1979.
Zedric, Lance Q., and Michael F. Dilley. "Raid on Oransbari." *Behind the Lines*, November–December 1995.

Interviews

Kittleson, Galen C., SGM (ret.). Telephone interview, 1 October 1996.
Lupyak, Joseph W., SGM (ret.). Telephone interview, 30 September 1996.

APPENDIX A

Origins of American Special Units

At the time of early conflict between the colonists and the American Indians (1622), there were no standing armies in the colonies. Landowners were responsible for their own defense, which usually meant having to arm their family and employees to protect their property and, obviously, to provide an alert force. Armed men who patrolled in search of an enemy force were said to be "ranging." From this came the habit of referring to the individual men as "Rangers." The earliest record of such an individual comes from a letter written by Captain John Smith of Jamestown in 1622. Smith mentions a "Captaine Butler" who was "… ranging in his Boat [sic] …" Between 1622 and 1648, more landowners began to designate employees to "range the forests and streams."

In 1634 there is a record of a Virginia landowner named Captain William Claiborne appointing one of his employees, Edward Backler (or Blackler) as a "rainger" for his Maryland plantation on Kent Island, in Chesapeake Bay. Backler served as a Ranger for two years, patrolling Claiborne's trading post areas to provide warnings of potential attacks by the Wicomesse Indians in the area. It is likely that other landowners also used employees in a similar fashion. By 1648, however, the era of privately hired Rangers had come to an end. In that year, several colonial governments hired parties of Rangers to guard the colony's frontiers. From that point until 1670, when King Philip's War broke out in Rhode Island colony, the use of these parties of Rangers was sporadic and records are almost non-existent about their formation and membership. It is at this point that the outline contained in the appendix picks up and continues to the present day. The units from 1670 until the Revolutionary War were part of the British Army.

Part 1. Pre-Revolution units[1]

Period	Unit/Leader	Conflict	Where/how employed
1670–76	Willets Rangers (Captain Thomas Willet)	King Philip's War	Bristol, Rhode Island; scouting force against Wampanoag (aka Pokanoket) Indians.
1675–1704	Church's Rangers (Captain Benjamin Church[2])	King Philip's War Queen Anne's War	New England—punitive raids against American Indians and French in Port Royal and Quebec.
1679–1774 (est.)	Rangers	French and Indian War (Seven Years' War) Dunmore's War	Virginia—scouted American Indians; fought with local militia; fought with small Ranger units from Maryland, North Carolina, and Pennsylvania.
1721–24	Moulton's Rangers[3] (Captain Jeremiah Moulton)		Maine—raids against Norridgewock Indians.
1732–47	English & Highland Rangers[4]	War of Jenkins' Ear King George's War	Georgia—raised by Governor James Ogelthorpe, fought against Spanish in Georgia and Florida.
1743	Rangers; Southern Scouts[5]	King George's War	South Carolina—fought Spanish in Florida with Georgia Rangers.

Period	Unit/Leader	Conflict	Where/how employed
1750–63/64	Gorham's Rangers[6] (Captain John Gorham[7])	French and Indian War (Seven Years' War)	Massachusetts—scouts against French in Nova Scotia; included at least three companies; later fought with Wolfe at Quebec and in an expedition against Havana.
1755–63	Royal American Rangers[8] (Lt. Col Henri Bouquet)	French and Indian War (Seven Years' War)	Four battalions trained in and employed 'ranging' tactics.
1756–60	Dunn's Rangers[9] (Captain Hezekia Dunn) Gardner's Rangers[10] (Captain Gardner)	French and Indian War (Seven Years' War)	New Jersey—patrolled the colony's western frontier against American Indian raids.
1756–63	Rogers Rangers[11] (Major Robert Rogers[12])	French and Indian War (Seven Years' War)	New York, Montreal, New England, Quebec, Detroit—scouting and raiding missions against the American Indians and French.
1762–63	Hopkins Rangers[13] (Captain Joseph Hopkins)	French and Indian War (Seven Years' War)	Fought at Detroit during Pontiac's siege.
1773–76	Georgia Rangers		Georgia—raised by Governor Sir James Ogelthorpe; patrolled colony's frontiers.

Part 2. American Revolutionary War units[14]

Period	Unit/Leader	Conflict	Where/how employed
1775–77	Marblehead Mariners[15] (Lt. Col. John Glover)	American Revolution	New York, Pennsylvania, and New Jersey—evacuated General George Washington's army from Long Island to Manhattan and later his troops across the Delaware River for its raid on Trenton.
1775–78	Morgan's Rangers[16] (Colonel Daniel Morgan)	American Revolution	Virginia and south
1775–80	South Carolina Rangers[17]		South Carolina
1776	Knowlton's Rangers[18] (Lt. Col. Thomas Knowlton)	American Revolution	Connecticut and New York
1776–81	Whitcomb's Rangers (Major Benjamin Whitcomb)	American Revolution	Two companies—conducted reconnaissance missions in New Hampshire and Vermont.
1778–81	Marion's Partisans (Brigadier General (BG) Francis Marion[19])	American Revolution	South Carolina
1778–81	(BG Thomas Sumter)	American Revolution	South Carolina
1778–81	(BG Andrew Pickens)	American Revolution	South Carolina

Part 3. Early American Wars

Period	Unit/Leader	Conflict	Where/how employed
1791–94	Wood Rangers (BG Anthony Wayne)	Northwest Indian Wars	Elements of Wayne's force received special "ranger tactics" training, defeated American Indians at battle of Fallen Timbers.
1794–1812	Rangers		Patrolled U.S. frontier to scout for the army.
1812–15	Rangers (Colonel William Russell) (Colonel John Coffee)	War of 1812	Seventeen companies raised to patrol the U.S. frontier from Louisiana to Michigan; scouted for General Andrew Jackson before the Battle of New Orleans; some units continued to patrol after the war.
1818	Rangers	First Seminole War	Two companies[20] scouted for General Andrew Jackson's campaign in Florida.
1831–33	Mounted Rangers (Major Henry Dodge)		Battalion (six companies) incorporated into regular army; patrolled U.S. frontier.
1835–90	Texas Rangers	Texas War for Independence; Mexican War; Indian Wars.	Acted as scouts and guides for military operations.

Part 4. U.S. Civil War[21]

Units of the U.S. Army

Period	Unit/Leader	Conflict	Where/how employed
1862	Andrews Raiders (James J. Andrews)[22]	Civil War	Georgia—captured a train at Big Shanty; destroyed rail and bridges and cut telegraph lines north before abandoning the mission north of Ringgold.
1862–65	Loudon Rangers[23] (Captain Samuel Means)	Civil War	Virginia, West Virginia, Maryland, and Pennsylvania—conducted raiding and scouting operations against the Army of Northern Virginia and fought against Mosby's Rangers.

Units of the Confederate Army

Period	Unit/Leader		Where/how employed
1861–63	Morgan's Raiders[24] (BG John H. Morgan)[25]		Kentucky, Tennessee, Indiana, Ohio, and West Virginia—conducted raiding operations.
1862–65	McNeill's Rangers[26] (Captain John McNeill)[27]		Virginia and Maryland—conducted raiding operations.
1862–65	Mosby's Rangers[28] (Colonel John S. Mosby)		Virginia—conducted raiding operations.

Part 5. Post U.S. Civil War

Period	Unit/Leader	Conflict	Where/how employed
1866–92	Indian Scouts	Indian Wars	American West—conducted scouting and reconnaissance missions in support of army cavalry and infantry units.
1898	Rough Riders[29] (Colonel Leonard Wood)	Spanish–American War	Cuba

Part 6. World War I

Period	Unit/Leader	Conflict	Where/how employed
1917–present	Submarines[30]	World War I and every war since.	Oceans of the world—originally operated in the Atlantic Ocean—expanded to worldwide missions after the war; conducted raids on shipping, specialized intelligence operations, and reconnaissance missions, including inserting special units, spies, and saboteurs.
1917–18	Indian Code Talkers	World War I	Europe—U.S. Army signal units used Choctaw Indians to speak their native language in radio communications instead of having to encode and decode messages.

Part 7. World War II and immediate post-war

Period	Unit/Leader	Conflict	Where/how employed
1940–present	Airborne forces—Army[31]	World War II and every war since	North African, Italian, European, and the Pacific Theaters and worldwide after the end of World War II—parachute assaults prior to main force landings, raids, and reinforcing actions.
1940–44	Airborne forces—Marines[32]	World War II	Pacific Theater—missions planned but never executed; misused in conventional infantry roles.
1941–45	Office of Strategic Services[33] (MG William Donovan)	World War II	All theaters except Southwest Pacific—conducted intelligence gathering, aid to resistance groups, counterintelligence operations, psychological operations, and liaison with other agencies and governments.
1941–present	Alaska Scouts[34]	World War II and every war since	Alaska—security operations in the Aleutians.

ORIGINS OF AMERICAN SPECIAL UNITS • 133

Period	Unit/Leader	Conflict	Where/how employed
1941–44	Navy Scouts & Raiders[35] (Commander Phil Buckles)	World War II	North African, Mediterranean, European, and Pacific Theaters—cut anti-shipping obstacles, guided landing forces ashore, conducted beach reconnaissance prior to amphibious landings, acted as Beach Masters.
1942–present	Marine Force Recon[36]	World War II and every war since	Pacific Theater initially, worldwide missions after the war—conducted raiding and reconnaissance missions.
1942	Doolittle's Raiders[37] (Lt. Col. James Doolittle)	World War II	Japan—conducted one direct action mission by bombing Japanese mainland.
1942–present	Psychological Operations[38]	World War II and every war since	European Theater during the war and expanded to worldwide mission since—conducted black and white propaganda missions.
1942–44	Marine Raiders[39] (Lt. Col. Merritt Edson) (Lt. Col. Evans Carlson)	World War II	Pacific Theater—conducted hit-and-run amphibious raiding missions.

Period	Unit/Leader	Conflict	Where/how employed
1942–44	First Special Service Force[40] (Colonel Robert T. Frederick)	World War II	Alaska, Italy, and Southern France—prepared for airborne insertion in Kiska, Alaska, used properly in mountains.
1942–44	Darby's Rangers[41] (Lt. Col. William O. Darby)	World War II	North Africa and Italy—conducted beachhead reconnaissance, combat patrols and raids, misused on last mission as conventional infantry.
1942–44	Jedburghs[42]	World War II	France, the Netherlands, and Belgium; some operated from North Africa and onto the European continent—coordinated and supported resistance organizations.
1942–43	29th Ranger Battalion[43]	World War II	Europe—conducted several raids.
1942–45	Navy Group China[44] (Captain Milton Miles)	World War II	China—provided weathermen in Mongolia, a carrier pigeon organization, trained Chinese counterespionage agents and guerrilla units including river raiders.
1942–45	OSS Operations Groups[45]	World War II	European Theater—trained and fought with resistance groups.

Period	Unit/Leader	Conflict	Where/how employed
1942–45	Allied Intelligence Bureau[46]	World War II	Southwest Pacific—oversaw clandestine intelligence collection operations behind enemy lines, supported guerrilla campaigns, and sent trained saboteurs into enemy controlled areas.
1942–45	Navajo Code Talkers[47]	World War II	Pacific Theater—conducted signal communications in code based on Navajo language.
1942–44	99th Infantry Battalion[48]	World War II	Europe—misused as a conventional infantry unit.
1942–45	Airborne Troop Carriers	World War II	North Africa, Europe—dropped paratroopers, tugged gliders, and resupplied these troops after inserting them for their missions.
1942–44	Naval Combat Demolitions Units[49]	World War II	Atlantic, Mediterranean, and European Theaters—conducted explosive ordnance disposal, beach obstacle demolitions, and hydrographic reconnaissance prior to amphibious assaults.

Period	Unit/Leader	Conflict	Where/how employed
1943–present	Civil Affairs[50]	World War II and every war since	All theaters during the war; expanded to worldwide mission after the war—worked with local governments in occupied areas.
1943–45	Amphibious Scouts[51] (Commander William Coultas)	World War II	Pacific—conducted beachhead reconnaissance for future amphibious operations.
1943–45	2nd and 5th Ranger Battalions[52]	World War II	Europe—beginning with D-Day invasion, conducted special missions and were occasionally misused in conventional infantry roles.
1943–45	Alamo Scouts[53]	World War II	New Guinea, Leyte, Philippines, and Japan—conducted intelligence reconnaissance missions, raids on POW camps, and trained and fought with guerrilla units in the Philippines; conducted personal security missions and training later classes at the Alamo Scouts Training Center when not in combat.

Period	Unit/Leader	Conflict	Where/how employed
1943–45 1985–present	10th Mountain Division[54]	World War II and every war since 1985	Italy in World War II, later given worldwide mission—conducted operations in mountainous areas
1943–44	Merrill's Marauders[55] (BG Frank Merrill[56])	World War II	Burma—conducted intelligence and raid missions behind Japanese lines.
1943–74	Underwater Demolition Teams[57]	World War II, Korea, and Vietnam	Pacific during the war, then given a worldwide mission—removed natural and man-made obstacles and demolitions, scouted beaches, and guided invasion forces ashore.
1944–45	Carpetbaggers[58]	World War II	Europe—used by OSS and SOE to insert agents and equipment into enemy territory, including Jedburgh teams, OSS Ops Groups, and, later, for dropping propaganda leaflets.
1944–45	Air Commandos[59] (Lt. Col. Philip Cochran) (Lt. Col. John Alison)	World War II	Burma and China—tugged gliders for troop insertion, resupplied units on ground, flew rescue missions for downed air crews.

Period	Unit/Leader	Conflict	Where/how employed
1944–45	6th Ranger Battalion[60] (Lt. Col. Henry Mucci)	World War II	Philippines—conducted security missions with navy Amphibious Scouts and a POW camp raid with Alamo Scouts, and combat patrols thereafter.
1945	SAARF[61] (Brigadier A. J. Nichols)	World War II	Europe—parachuted into area near German POW camps prior to them being overrun by friendly forces to observe, report, or intercede if necessary to protect the lives of the POWs.
1946–present	Air Force Pararescuemen (PJs)[62]	Every war since formation	Worldwide mission—saving the lives of air crew who were involved in aircraft disasters, accidents, ditching, crash landings, abandonments, and supported NASA astronaut recovery missions.

ORIGINS OF AMERICAN SPECIAL UNITS • 139

Part 8. Korea and post-Korea

Period	Unit/Leader	Conflict	Where/how employed
1950–51	Ranger Airborne Companies[63]	Korean War	Korea—conducted combat patrols, raids on key facilities and equipment concentrations in enemy rear areas.
1951–53	8240th Army Unit[64] (Colonel John McGee)	Korean War	Korea—organized, trained, led, and assisted in the insertion of guerrillas and partisan forces behind enemy lines.
1951–present	Air Commandos[65]	Korean War and every war since	Korea then later received worldwide mission—psychological warfare, agent insertion, air rescue operations.
1952–present	Special Forces[66]	Every war since formation	Worldwide missions—organizing, training, and leading counterinsurgency units of foreign nations, conducting direct action missions and prisoner rescue operations.
1953–present	Combat Control Teams[67]	Every war since formation	Worldwide missions—provide expert air support and communications capabilities; conduct special reconnaissance, establish assault zones or airfields.

Part 9. Vietnam and post-Vietnam

Period	Unit/Leader	Conflict	Where/how employed
1956–74	Long Range Reconnaissance Patrols[68]	Vietnam War	Worldwide mission while they existed—conducted combat patrols and some direct-action missions for their supported divisions and commands.
1962–present	SEAL Teams[69]	Every war since formation	Worldwide missions—beachhead reconnaissance, explosive ordnance disposal, training foreign special navy units, conducting direct action missions including prisoner rescue, demolitions operations.
1974–present	75th Ranger Regiment[70]	Every war since formation	Worldwide missions—combat patrolling and direct-action missions.
1980–present	Night Stalkers[71]	Every war since formation	Worldwide missions—specialized air support and transport for Special Forces and Ranger units.

Notes

1. These units were raised by the British Army and included both British regular soldiers and colonists.
2. A member of Willet's Rangers during King Philip's War.
3. Included companies led by Captains Bourne, Bare, and John Harmon.
4. Also known as the Georgia Rangers.
5. Also known as the Florida Rangers.
6. Also known as Nova Scotia Ranging Company, the Corps of Rangers, and North American Rangers.
7. Also spelled as Goreham. Company commanders included Joseph Gorham (brother) and the brothers Isaac and Israel Putnam.
8. Also known as the 60th Foot (Royal American) Regiment.

9   Also known as the New Jersey Ranging Company, part of the Frontier Guard.
10 Also part of the Frontier Guard.
11 Almost 20 companies were raised but only nine were active at any one time. Subordinate commanders included Richard Rogers (brother), John Stark, Israel Putnam, and Joseph Gorham.
12 During the American Revolutionary War, Rogers raised the Queen's Rangers or Queen's American Rangers for the British, later called Simcoe's Rangers after a subordinate commander. Although this unit served with distinction, Rogers was relieved of command after the first engagement.
13 Also known as the Queen's Rangers or the Queen's Royal American Rangers.
14 These do not include units loyal to the king, such as the Queen's Rangers (see footnote for Major Robert Rogers), the King's (Carolina) Rangers, or the six companies known as Butler's Rangers (Colonel Walter Butler) which conducted raids in the Mohawk Valley between 1777 and 1781. Some of the operations of Butler's Rangers were as far away as Detroit and South Carolina.
15 Also known as the 14th Continental Regiment. Although this was an army unit, it is included here for its two special boat/amphibious operations.
16 Also known as Morgan's Riflemen.
17 Also known as the 3rd South Carolina Regiment.
18 Although called Rangers, this was a light infantry unit not generally used to scout, raid, or conduct reconnaissance missions. Its training included some of the tactics developed and refined by Bouquet and Rogers and it was used mostly to gather intelligence. Knowlton was killed at the Battle of Harlem Heights in September 1776 and the unit was disbanded.
19 Marion was known as the "Swamp Fox."
20 Commanded by Captain Boyle and Captain McGirt.
21 During this war, there were more than 400 units that included the term "Rangers" in their names. Many of these were guerrilla organizations not formally units of either army; most of them were organized under the Partisan Ranger Act passed by the Confederate Congress on 27 March 1862. It brought local fame to such leaders as Turner Ashby, E. V. White, John B. Imboden, Harry Gilmor, Kincheloe, O'Farrell, and others. This Act was repealed in April 1864. The repeal applied to all units organized under the act except those of McNeill and Mosby. Many of these units did not employ "ranger tactics" nor did they receive any "ranger training," the major measures for inclusion in this listing. The units which follow, by their operations, should be included.
22 Andrews was a civilian; 14 of the 18 raiders were the first to be awarded the Medal of Honor. Eventually all but two received this award. Andrews was not included because he was not in the military.
23 Originally raised as the Loudon Independent Guards. This unit varied in size but at its peak included four companies. Company commanders included Captains Michael Mullen and James Grubb. After Means resigned in April 1864, the unit was commanded by Captain Daniel Keyes and, later, Lieutenant Edwin Grover.

24 Officially known as the 2nd Kentucky Cavalry Regiment. Subordinate commanders included Basil Duke, James West, and Van Buren Sellers.
25 Morgan was captured in July 1864 and later escaped. He raised another unit in April 1864, but it did not last long. Morgan was killed by a Union sniper.
26 Originally mustered into service as Company I, 1st Regiment Virginia Partisan Rangers, later designated Company E, 18th Virginia Cavalry. In early 1863, it broke away from the 18th and operated independently thereafter.
27 McNeill was killed in action in November 1864; he was succeeded in command by his son, Jesse.
28 Officially known as the 43rd Virginia Cavalry.
29 Officially known as the 1st United States Volunteer Cavalry, it is included in this listing because of its recruitment practices. It was employed in conventional cavalry roles.
30 Although the first U.S. Navy submarine, *Holland (SS-1)*, was officially commissioned on 11 April 1900, it was not until World War I that operational missions were assigned. Initially, most of these missions consisted of convoy escort duties. Missions expanded greatly during and after World War II.
31 Plans to parachute the 1st Infantry Division near Metz, Germany, were finalized in 1918 but the war ended before the plans were executed. Prior to the outbreak of World War II, the U.S. Army began testing the feasibility of using airborne troops. Units were formed prior to the beginning of the war and were eventually deployed in most major theaters. In Europe, this included one airborne corps (XVIII) consisting of three airborne divisions (82nd, 101st, and 17th, although the 82nd and 101st saw action and made combat jumps prior to the corps' activation), one other airborne division (13th), one airborne brigade (2nd Airborne Brigade, consisting of the 507th PIR and 508th PIR, which were attached to the 82nd Airborne Division during D-Day operations), one separate parachute regiment (517th PIR), two separate parachute battalions (509th PIB and 551st PIB), and one separate glider regiment (550th AIB). In the Pacific, it included one airborne division (11th) and one separate parachute regiment (503rd PIR). In the United States, it included one separate parachute battalion (555th PIB) and two separate parachute battalions for training (541st PIB and 542nd PIB). The airborne divisions included both parachute and glider regiments. After the war, the 13th and 17th divisions, the airborne brigade, all the separate parachute regiments, all the separate parachute battalions except the 509th, the separate glider battalion, and one separate training battalion were deactivated. Eventually all glider units were disbanded or reorganized as parachute units. A training battalion remained active but unnumbered until the advent of the regimental system. Over the years after World War II, one airborne regiment was activated and deployed during Korea, reinforced to brigade size and used in Vietnam, and several other conflicts under various numerical designations (e.g., 187th ARCT, 173rd Airborne Brigade). There were several airborne divisions in the U.S. Army Reserve and one separate airborne regiment in the National Guard; all were eventually taken off jump

status. Following the Vietnam War, the airborne corps and one of the airborne divisions were taken off jump status and converted to air assault (the 101st), and the one remaining airborne division (11th) was redesignated as straight leg infantry. Active units currently include the 82nd Airborne Division, the 173rd Airborne Brigade, and the 509th Airborne Battalion. On 6 June 2022, the 11th Airborne Division was reactivated and stationed in Alaska, where it is currently nicknamed the Arctic Angels. In addition to being on jump status, members of the unit are trained to operate in very cold weather conditions.

32 In the Pacific, this included three separate parachute battalions (1st MPB, 2nd MPB, and 3rd MPB) that were eventually consolidated into one parachute regiment (1st MPR). A separate parachute battalion (4th MPB) was used for training. Units never conducted any combat jumps.

33 Originally activated as the Coordinator of Information, the OSS was formalized in June 1942 and deactivated in December 1945, with its functions split between the War Department and the Department of State.

34 Also known as Alaska Defense Command Scout Detachment (Provisional) and Castner's Cutthroats. The Scouts were trained to operate in three- to five-man teams and to infiltrate by rubber boat, submarine, PBY/flying boats, native *bidarkas*, and various commercial craft. The unit never exceeded a total strength of 68. In 1947, the Scouts were converted to a National Guard unit.

35 Unit members were later absorbed by Underwater Demolition Teams.

36 Originally known as the Marine Amphibious Reconnaissance Company, they were formed to provide Marine divisions with an elite reconnaissance squad. Following the war, this effort expanded its training and selection process.

37 Included in this listing because the unit flew U.S. Army bombers from a navy aircraft carrier (USS *Hornet*). Planning for the bombing of the Japanese home islands began in February 1942. Shortly thereafter, volunteers from the 17th Bombardment Group conducted training on their B-25s at Eglin Field, Florida.

38 Originally created as the Psychological Warfare Branch of the U.S. Army Staff. Psychological warfare groups were created and placed under the operational control of the OSS. In 1950, the Psychological Warfare Center was established at Fort Bragg, North Carolina. Psychological operations units are now part of army special operations and units are in the active army and the U.S. Army Reserve.

39 In the Pacific, this included four raider battalions (1st MRB, 2nd MRB, 3rd MRB, and 4th MRB) which were eventually consolidated into one raider regiment (1st MRR). In September 1943, the 1st MRB and 4th MRB were separated from the 1st MRR (which was deactivated) because of injuries; the 2nd MRB and 3rd MRB were transferred to the 2nd MRR. Commanders for the 1st MRB and 2nd MRB were Colonel Merritt Edson and Lieutenant Colonel Evans Carlson. There were several commanders for the other raider units.

40 Also known as the Devil's Brigade or the Black Devils. This was a multi-national unit composed of U.S. and Canadian troops; the soldiers received parachute, ski,

mountain, and severe weather training. Initially raised to use a sled-type vehicle in the snow in Norway, this requirement was later dropped.

41 The 1st Battalion was formed to conduct hit-and-run raids and collect intelligence. Following the Allies' invasion of French North Africa, Darby was authorized to recruit two additional battalions, the 3rd and the 4th. The 1st and 3rd Battalions were destroyed at Cisterna, Italy, and those remaining in the 4th Battalion were transferred to the First Special Service Force.

42 A multi-national effort, conducted in conjunction with the British SOE, to use three-man teams to train, supply, and support resistance groups in France, the Netherlands, and Belgium. Planning for this organization began in July 1942, with recruiting drives starting almost immediately. Final approval for the operations by OSS and SOE came in December 1943. Following Operation *Market Garden*, all teams were withdrawn and the resources re-allocated as needed. At least one member of each team was of the nationality of the country the team operated in.

43 Some of Darby's original Rangers and volunteers from within the 29th Infantry Division formed the 29th Ranger Battalion (Provisional). This unit conducted several raids with British commandos but was deactivated when additional separate Ranger Battalions (2nd, 3rd, 4th, and 5th) were formed.

44 Also known as the Rice Paddy Navy, this unit worked in coordination with SACO, the Sino-American Cooperative Organization. Miles was later promoted to rear admiral while in command of this unit.

45 Original members recruited from 122nd Infantry Battalion, a Greek-speaking unit. Individual operations groups were often only about platoon sized and most members in the unit were proficient in the language of the unit's targeted country.

46 Established as a staff element under General Douglas MacArthur. MacArthur refused to let the OSS operate within his theater since it was not under his direct control and did not report to him. This organization oversaw special operations within the Southwest Pacific Theater and included both U.S. and Australian officers. Coordinated and supported the operations of Philippine guerrilla units led by stay-behind and especially inserted American officers.

47 Organized by the U.S. Marine Corps to provide secure radio communications. The Navajo language is an oral language, hence no dictionary exists. A special code was developed, based on the language, which had to be memorized by Navajo Marines before deploying overseas.

48 Raised to be composed of Norwegian speakers, unit members received specialized ski and mountain training at Camp Hale, Colorado, but were misused in conventional infantry roles and never assigned to Norway. The battalion was eventually merged into the 474th Infantry Regiment with American members of the First Special Service Force and the remainder of Darby's Rangers.

49 Initially an army-navy organization, it later became all navy. At its greatest strength it was composed of 12 teams. When deactivated after D-Day, many members transferred to the Underwater Demolition Teams.

50 Originally created to rebuild local governments in occupied countries by establishing military government teams. Following the creation of Special Forces, this capability was teamed with psychological operations units under the control of the Army Special Operations Command.
51 Also known as the U.S. Seventh Fleet Special Service Unit #1, this was initially a multi-national (American/Australian) and multi-service (army, navy) organization that later only belonged to the U.S. Navy. The unit worked with Alamo Scouts and 6th Ranger Battalion on some missions.
52 Raised and trained in the United States.
53 Also known as the Sixth Army Special Reconnaissance Unit, it was created by General Walter Krueger, the U.S. Sixth Army commander in New Guinea. Alamo Scouts Training Center conducted training for the teams. Graduates not sent to teams were returned to their parent Sixth Army units to provide training to parent unit members and conduct reconnaissance missions. Deactivated in Japan.
54 Originally activated as the 10th Light Division (Alpine) and trained in ski, mountain, and severe weather operations, it was redesignated the 10th Mountain Division in 1944. Following the war, it was again redesignated as the 10th Infantry Division and used as a training unit, then as a full combat division. It was deactivated in 1958 and reactivated in 1985 as the 10th Mountain Division (Light Infantry) and stationed at Fort Drum, New York.
55 Also known by the official code-name of GALAHAD and officially as the 5307th Composite Unit (Provisional). Misused in conventional infantry role following the capture of an airfield at Myitkyina, the unit was later redesignated as the 475th Infantry Regiment and included in the Mars Task Force in a conventional infantry role.
56 Replaced by Colonel Charles Hunter following Merrill's medevac after a second heart attack.
57 Also known as Navy Frogmen.
58 Also known as the 801st Bombardment Group (Heavy), and later as the 492nd Bombardment Group (Heavy) of the U.S. Eighth Air Force. The group's B-24s were modified for optimum stealth since their secret missions were flown at night over enemy territory. After September 1944, they flew C-47s. As the war progressed, the unit moved from airfields in England to France.
59 Also known as the 5318th Provisional Unit (Air) and later redesignated as the 1st Air Commando Group. The 2nd Air Commando Group and the 3rd Air Commando Group were activated later. The unit initially operated out of India to work with Brigadier Orde Wingate's Chindits. Tactics developed by the unit are still used by the U.S. Air Force Special Operations Command.
60 Redesignated from the 98th Field Artillery Battalion, the battalion trained in New Guinea before its first missions.
61 Officially known as the Special Allied Airborne Reconnaissance Force, this was a multi-national unit. It was composed of paratroopers from the U.S., Britain, Belgium, France, and Poland who had formerly served in the OSS, airborne

units, and SOE. Some former SOE/OSS female agents and former Jedburgh Team members also were in this unit. The basic unit was a three-person team. The only operational airborne mission was the jump of six teams in near the POW complex at Altengrabow in late April 1945. Disbanded in July 1945.

62 Initially formed by the U.S. Army Air Force as the Air Rescue Service, the mission later belonged to the USAF. Pararescue units are part of the U.S. Air Force Special Operations Command and the Air Combat Command, depending on the specific rescue unit.

63 Eighteen companies were formed, 17 of which were airborne. Those companies not in Korea served in the United States and with units in Europe. Only eight companies saw combat. The 2nd Ranger Airborne Company and 4th Ranger Airborne Company, along with 2nd and 3rd Battalions of the 187th Airborne Regimental Combat Team conducted an airborne assault behind North Korean lines. The activation of the Ranger Airborne Companies led to the establishment of the Ranger training course at Fort Benning, Georgia, which continues to the present. A hallmark of this training is that graduates are expected to return to their units (if not selected to serve with Ranger units) and pass on their training to unit members, similar to the mission of the Alamo Scouts.

64 Also known as the Attrition Warfare Section, Eighth Army G-3 Miscellaneous Division, then redesignated as the 8086th Army Unit, and still later redesignated as the Far East Command Liaison Detachment Korea, 8240th Army Unit. Raised organizations were designated as Partisan Infantry Regiments, then United Nations Partisan Forces, Korea, and still later as United Nations Partisan Infantry, Korea. At peak strength there were seven regiments numbering 21,000 members.

65 Officially known as the Air Force Special Operations Command, it assumed its nickname from the World War II units. Initially formed by Captain Harry Aderholt, these forces, at first, included units that flew missions for military organizations and the CIA. The current headquarters is at Hurlburt Air Force Base, Florida, with training sites at several locations including Lackland Air Force Base, San Antonio, Texas, and Kirtland Air Force Base, Albuquerque, New Mexico. Various Air Force aircraft are used to support the special operations units.

66 Also known as the Green Berets after their distinctive headgear, Special Forces draw their lineage from several units including the Jedburghs and the Alamo Scouts. The organization includes several Special Forces Groups in the active U.S. Army, U.S. Army Reserve, and the National Guard. During the Vietnam War, a Special Forces organization known by the cover name Studies and Observation Group (SOG) conducted cross-border operations in Cambodia, Laos, and North Vietnam on direct action, reconnaissance, and prisoner rescue missions. During the late 1970s a similar organization, known as Delta Force (officially 1st Special Forces Operational Detachment—Delta), was created to conduct raids and prisoner release missions. The basic unit of a Special Forces Group is called an A Team, composed of 12 soldiers all cross-trained in several specialties including languages, weapons, engineering, communications, intelligence, and medical.

67 This is a USAF organization, an outgrowth of the World War II Army Pathfinders. Members must be FAA-certified Air Traffic Controllers; many are also Joint Terminal Attack Controllers. Members are assigned individually to U.S. Army Airborne, Ranger, and Special Forces units, and U.S. Navy SEAL teams.

68 Initially formed by the 11th Airborne Division in Germany, this concept was spread to many of the divisions of the active U.S. Army and U.S. Army Reserve. In 1969, all LRRP units were redesignated as companies in the reactivated 75th Ranger Regiment, which consisted of these companies only, and not to be confused with the later formation with the same identification. The designation is from the 75th Regimental Combat Team, which drew its lineage from the 5307th Composite Unit (Provisional) (see footnote for Merrill's Marauders). Companies were deactivated from the regiment in 1974 and converted to Long Range Surveillance Units as part of their parent division.

69 Initially formed to expand the mission of Underwater Demolition Teams, the organization now includes 10 teams and supporting special boat units. Odd numbered teams are stationed in San Diego, California, under Navy Special Warfare Group One, while even numbered teams are stationed at Little Creek, Virginia, under Navy Special Warfare Group Two. SEAL Team Six, under the cover name of Development Group, has a similar mission to the U.S. Army's Delta Force.

70 Formed to bring back the Ranger concept of operations to the U.S. Army. The regimental headquarters and 1st Ranger Battalion are stationed at Fort Benning, the 2nd Ranger Battalion is stationed at Fort Stewart, and the 3rd Ranger Battalion is stationed at Joint Base Lewis-McChord, Washington. The regiment draws its lineage from various World War II and Korean War Ranger units which are not color-bearing. This lineage also includes the formerly active 75th Ranger Regiment, which is color bearing.

71 Originally formed as Task Force 158 because its early members and aircraft came from 158th Aviation Battalion, 101st Airborne Division (Air Assault), Fort Campbell, Kentucky. The unit was later redesignated as Task Force 160 and then renamed the 160th Special Operations Aviation Regiment. The 1st and 2nd Battalions are stationed at Fort Campbell to support Special Forces operations, and the 3rd Battalion is stationed at Hunter Army Airfield near Fort Stewart, Georgia, where it supports the 75th Ranger Regiment.

APPENDIX B

Organization of the Ranger Battalions in World War II

Battalion	Place Activated	Place Inactivated
1st Ranger Bn	Carrickfergus, Northern Ireland	Anzio, Italy
2nd Ranger Bn	Camp Forest, TN	Camp Patrick Henry, VA
3rd Ranger Bn	Nemours, Morocco	Anzio, Italy
4th Ranger Bn	North Africa	Camp Butner, NC
5th Ranger Bn	Camp Forest, TN	Camp Myles Standish, MA
6th Ranger Bn	58th Field Arty Bn, Ft Lewis, WA; Redesignated: 25 September 1944	Japan

APPENDIX C

Organization of United States Air Force Pararescue units

Unit	Location
347th Rescue Gp	Moody AFB, GA
38th Rescue Sqdn	Moody AFB, GA
41st Rescue Sqdn	Moody AFB, GA
71st Rescue Sqdn	Moody AFB, GA
563rd Rescue Gp	Davis-Monthan AFB, AZ
48th Rescue Sqdn	Davis-Monthan AFB, AZ
55th Rescue Sqdn	Davis-Monthan AFB, AZ
58th Rescue Sqdn	Nellis AFB, NV
66th Rescue Sqdn	Nellis AFB, NV
79th Rescue Sqdn	Davis-Monthan AFB, AZ
31st Rescue Sqdn	Kadena AFB, Okinawa, Japan
33rd Rescue Sqdn	Kadena AFB, Okinawa, Japan
56th Rescue Sqdn	RAF Lakenheath, Suffolk, England, UK

Unit	Location
106th Rescue Wing (NYAFNG)	Francis S. Gabreski Airport
101st Rescue Sqdn (NYAFNG)	Francis S. Gabreski Airport
102nd Rescue Sqdn (NYAFNG)	Francis S. Gabreski Airport
103rd Rescue Sqdn (NYAFNG)	Francis S. Gabreski Airport
129th Rescue Wing (CAAFNG)	Moffett Federal Airfield
129th Rescue Sqdn (CAAFNG)	Moffett Federal Airfield
130th Rescue Sqdn (CAAFNG)	Moffett Federal Airfield
131st Rescue Sqdn (CAAFNG)	Moffett Federal Airfield
210th Rescue Sqdn (AKAFNG)	Joint Base Elmendorf-Richardson
210th Rescue Sqdn, Det 1 (AKAFNG)	Eielson AFB
211th Rescue Sqdn (AKAFNG)	Joint Base Elmendorf-Richardson
212th Rescue Sqdn (AKAFNG)	Joint Base Elmendorf-Richardson
920th Rescue Wing (AFR)	Patrick AFB, FL
920th Ops Gp (AFR)	Patrick AFB, FL
39th Rescue Sqdn (AFR)	Patrick AFB, FL
301st Rescue Sqdn (AFR)	Patrick AFB, FL
308th Rescue Sqdn (AFR)	Patrick AFB, FL
943rd Rescue Gp (AFR)	Davis-Monthan AFB, AZ
305th Rescue Sqdn (AFR)	Davis-Monthan AFB, AZ
306th Rescue Sqdn (AFR)	Davis-Monthan AFB, AZ
304th Rescue Sqdn (AFR)	Portland International Airport

NOTE: bold units are parent organization for subordinate units at same location, in most cases.

APPENDIX D

Organization of the Ranger Companies in the Korean War

Company	Unit Attached
Eighth Army Ranger Co (8213 AU)	25th Inf Div, IX US Army Corps
GHQ Raider Co (8227 AU), X Corps Raider Co (8245 AU)	Special Activities Gp, X Corps
1st Ranger Co (Abn)	2nd Inf Div
2nd Ranger Co (Abn)	7th Inf Div, 187th Abn RCT
3rd Ranger Co (Abn)	Ranger Training Cmd, 3rd Inf Div, I Corps
4th Ranger Co (Airborne)	1st Cav Div, 187th Abn RCT, 1st Marine Div
5th Ranger Co (Abn)	25th Inf Div, I Corps
6th Ranger Co (Abn)	1st Inf Div
7th Ranger Co (Abn)	Ranger Training Cmd
8th Ranger Co (Abn)	24th Inf Div, IX Corps
9th Ranger Co (Abn)	31st Inf Div
10th Ranger Co (Abn)	45th Inf Div
11th Ranger Co (Abn)	40th Inf Div
12th Ranger Co (Abn)	28th Inf Div
13th Ranger Co (Abn)	43rd Inf Div
14th Ranger Co (Abn)	4th Inf Div

Company	Unit Attached
15th Ranger Co (Abn)	47th Inf Div
Ranger Training Cmd	Infantry School, Fort Benning, Georgia
Ranger Training Co A	Ranger Training Cmd
Ranger Training Co B	Ranger Training Cmd

APPENDIX E

Organization of the Special Forces Groups

Group/Location	Area of Responsibility
Training Group[1]	
1st SF Group, Joint Base Lewis-McChord, WA[2]	Pacific Region
3rd SF Group, Ft Bragg, NC	Sub-Saharan Africa except the Eastern Horn of Africa
5th SF Group, Ft Campbell, KY	Middle East, Persian Gulf, Central Africa, and Horn of Africa
6th SF Group	Southwest and Southeast Asia (Inactive)
7th SF Group, Eglin Air Force Base, FL[3]	North and South America, Central America, Caribbean
8th SF Group	Training armies of Latin America
10th SF Group, Ft Carson, CO[4]	Europe
11th SF Group	(Inactive)
12th SF Group	(Inactive)
19th SF Group, Draper, UT	ARNG
20th SF Group, Birmingham, AL	ARNG
38th SF Co, Anchorage, AK[5]	ARNG
46th SF Co, Lop Buri, Thailand	Thailand

Group/Location	Area of Responsibility
SF Operational Det Korea[6]	Korea
1st SF Operational Det—Delta	Worldwide

Notes

1. In 1969, incorporated into the U.S. Army Institute for Military Assistance.
2. 1st Bn was forward deployed to Torii Station, Okinawa.
3. Originally activated as the 77th SF Group on 16 September 1953; redesignated on 20 May 1960.
4. The first SF Group to be activated, in June 1952. The 1st Bn was forward deployed to Stuttgart, Germany.
5. Originally activated as the 38th SF Detachment.
6. Activated at Seong-Nam, Korea; assigned to the 1st SF Group in 1986.

APPENDIX F

Organization of the LRRP Companies, pre-Vietnam War

Company/Unit	Unit/Post Attached
110th Aviation Co (Surveillance), Airborne Reconnaissance Plt	US Army Southern European Task Force, Italy
V Corps LRRP units	V Corps, Federal Republic of Germany
Surveillance Plts (Provisional)	Seventh Army, Federal Republic of Germany
LRRP (Provisional)[1]	Seventh Army, Federal Republic of Germany
Recon Unit[2]	V Corps, Federal Republic of Germany
Recon Unit	VII Corps, Federal Republic of Germany
LRRP Co (Abn)	V Corps, Federal Republic of Germany
LRRP Co	VII Corps, Federal Republic of Germany
3rd Inf Div LRRP	3rd Inf Div, Federal Republic of Germany

Notes

1. These units were based on A Co, 2nd Bn, 51st Armor Inf Rgt, 4th Armor Div.
2. These units had a strength of 80 men.

APPENDIX G

Organization of the LRP Companies in the Vietnam War

Company/Unit	Unit/Post Attached
Co. D, 17th Inf (Abn) (LRP)	V Corps, Federal Republic of Germany; transferred to Ft Benning, GA, in 1968
Co. C, 58th Inf (Abn) (LRP)	VII Corps, Federal Republic of Germany; transferred to Ft Riley, KS, in 1968
Co. E, 20th Inf (LRP)	I Field Force, Vietnam
Co. F, 51st Inf (LRP)	II Field Force, Vietnam
Co. D, 151st Inf (LRP)	II Field Force, Vietnam
Co. E, 50th Inf (LRP)	9th Inf Div
Co. F, 50th Inf (LRP)	25th Inf Div
Co. E, 51st Inf (LRP)	196th Light Inf Bde; 23rd Inf (Americal) Division (Mechanized)
Co. E, 52nd Inf (LRP)	1st Cav Div (Airmobile)
Co. F, 52nd Inf (LRP)	1st Inf Div
Co. E, 58th Inf (LRP)	4th Inf Div
Co. F, 58th Inf (LRP)	101st Abn Div (Air Assault)
71st Inf Det (LRP)	199th Light Inf Bde
74th Inf Det (LRP)	173rd Abn Bde
78th Inf Det (LRP)	3rd Bde, 82nd Abn Div
79th Inf Det (LRP)	1st Bde, 5th Inf Div (Mechanized)

APPENDIX H

Organization of the Ranger Companies in the Vietnam War

Company	Unit/Post Attached
Co A (Ranger), 75th Inf	Ft Benning, Ft Hood
Co B (Ranger), 75th Inf	Ft Carson, Ft Lewis
Co C (Ranger), 75th Inf	I Field Force, Vietnam
Co D (Ranger), 151st Inf	II Field Force, Vietnam
Co D (Ranger), 75th Inf	II Field Force, Vietnam
Co E (Ranger), 75th Inf	9th Inf Div
Co F (Ranger), 75th Inf	25th Inf Div
Co G (Ranger), 75th Inf	196th Light Inf Bde; 23rd Inf (Americal) Div (Mechanized)
Co H (Ranger), 75th Inf	1st Cav Div (Airmobile)
Co I (Ranger), 75th Inf	1st Inf Div
Co K (Ranger), 75th Inf	4th Inf Div
Co L (Ranger), 75th Inf	101st Abn Div (Air Assault)
Co M (Ranger), 75th Inf	199th Light Inf Bde
Co N (Ranger), 75th Inf	173rd Abn Bde
Co O (Ranger), 75th Inf	3rd Bde, 82nd Abn Div
Co P (Ranger), 75th Inf	1st Bde, 5th Inf Div (Mechanized)

APPENDIX I

Organization of the 75th Ranger Regiment

Battalion	Location
Regimental HQ	Ft Benning, GA
1st Ranger Bn	Hunter Army Airfield, GA
2nd Ranger Bn	Joint Base Lewis-McChord, WA
3rd Ranger Bn	Ft Benning, GA
Regimental Special Troops Bn	Ft Benning, GA

APPENDIX J

Organization of Combat Control Teams

Squadron	Location
24th Spec Ops Wing	Hurlburt Air Force Base, FL
720th Spec Tactics Gp	Hurlburt Air Force Base, FL
21st Spec Tactics Sqdn	Pope Air Force Base, NC
22nd Spec Tactics Sqdn	Joint Base Lewis-McChord, WA
23rd Spec Tactics Sqdn	Hurlburt Air Force Base, FL
724th Spec Tactics Gp	Pope Air Force Base, NC
24th Spec Tactics Sqdn	Pope Air Force Base, NC
353rd Spec Operations Gp	Kadena Air Force Base, Okinawa, Japan
320th Spec Tactics Sqdn	Kadena Air Force Base, Okinawa, Japan
352nd Spec Operations Gp	RAF Mildenhall, United Kingdom
321st Spec Tactics Sqdn	RAF Mildenhall, United Kingdom
123rd Spec Tactics Sqdn (KYAFNG)	Louisville International Airport, KY
125th Spec Tactics Sqdn (ORAFNG)	Portland International Airport, OR

Note: bold units are parent organization for subordinate units at same location, in most cases.

APPENDIX K

Organization of the SEAL Teams

Team	Location
Naval Spec Warf Gp One	Naval Amphibious Base, Coronado, CA
Naval Spec Warf Gp Two	Naval Amphibious Base, Little Creek, VA
SEAL Team 1	Naval Amphibious Base, Coronado, CA
SEAL Team 2	Naval Amphibious Base, Little Creek, VA
SEAL Team 3	Naval Amphibious Base, Coronado, CA
SEAL Team 4	Naval Amphibious Base, Little Creek, VA
SEAL Team 5	Naval Amphibious Base, Coronado, CA
SEAL Team 6[1]	Dam Neck Annex, Virginia Beach, VA
SEAL Team 7	Naval Amphibious Base, Coronado, CA
SEAL Team 8	Naval Amphibious Base, Little Creek, VA
SEAL Team 10	Naval Amphibious Base, Little Creek, VA
SEAL Delivery Vehicle Team One	Pearl Harbor, HI
SEAL Delivery Vehicle Team Two[2]	Naval Amphibious Base, Little Creek, VA

Notes
1. Now known officially as Naval Special Warfare Development Group.
2. After de-activation, merged into SDVT-1.

APPENDIX L

Summary of U.S. Marine Corps Special Operations Organizations

MARSOC—USMC Spec Ops Cmd
Force Recon
Recon Bns
Scouts Snipers
Air and Naval Gunfire Liaison Cos (ANGLICO)
Radio Recon Teams (RRT)
Maritime Raid Force (MRF)
Maritime Special Purpose Force (MSPF)
USMC Special Reaction Team
Fleet Anti-Terrorism Security Team
Recapture Tactics Team

Bibliography

Adleman, Robert H., and George Walton. *The Devil's Brigade*. Philadelphia, PA: Chilton Books, 1966.
Author unknown. "Ranger Units' Lineage, Honors Go To 75th Ranger Regiment." *Army*, June 1986.
Author unknown. "Special Allied Airborne Reconnaissance Force." Wikipedia, no date.
Baker, Alan D. *Merrill's Marauders*. New York: Ballantine Books, 1972.
Bergen, Howard R. *History of 99th Infantry Battalion, U.S. Army*. Oslo: Emil Moestue A-S, 1945.
Biggs, Bradley. *The Triple Nickels: America's First All-black Paratroop Unit*. Hamden, CT: Archon Books, 1986.
Black, Robert W. *Rangers in Korea*. New York: Ivy Books, 1989.
Black, Robert W. *Rangers in World War II*. New York: Ivy Books, 1992.
Blair, Jr., Clay. *Ridgway's Paratroopers: The American Airborne in WWII*. New York: Doubleday, 1985.
Bowers, Everett P. "Expert POW Liberator." *Military History*, October 1996.
Bowers, Frank R., B. Hugh Tovar, and Richard Schultz, eds. *Special Operations in U.S. Strategy*. Washington, D.C.: National Defense University Press, 1984.
Bradley, Francis X., and H. Glen Wood. *Paratrooper*. Harrisburg, PA: Stackpole Books, 1962.
Brereton, Lewis H. *The Brereton Diaries: The War in the Air in the Pacific, Middle East and Europe, 3 October 1941–8 May 1945*. New York: William Morrow and Company, 1946.
Breuer, William B. *Geronimo! American Paratroopers in WWII*. New York: St. Martin Press, 1989.
Brown, Ashley, and Jonathan Reed, eds. *The Unique Units* (A volume in The Elite series). London: Orbis Publishing, 1989.
Bruning, John H. "The Last Jump: Task Force Gypsy at Aparri." Posted 2014. https://theamericanwarrior.com/2014/12/05/the-last-jump-task-force-gypsy-at-aparri/.
Burhans, Robert D. *The First Special Service Force: A War History of the North Americans 1942–1944*. Washington, D.C.: Infantry Journal Press, 1947.

Capp, Jimmy. "SAARF, Special Allied Airborne Reconnaissance Force." Posted 14 June 2009. https://www.usmilitariaforum.com.

Critchell, Laurence. *Four Stars in Hell*. New York: MacMillan, 1947.

Cuneo, John R. "The Early Days of the Queen's Rangers: August 1776 to February 1777." *Military Affairs*, Summer 1958, 65–74.

Cuneo, John R. *Robert Rogers of the Rangers*. New York: Oxford University Press, 1959.

David, Heather. *Operation: Rescue*. New York: Pinnacle Books, 1971.

De Trez, Michel. *First Airborne Task Force: Pictorial History of the Allied Paratroopers in the Invasion of Southern France*. Belgium: D-Day Publishing, 1998.

Devlin, Gerald M. *Silent Wings: The Saga of the U.S. Army and Marine Combat Glider Pilots of World War II*. New York: St. Martin's Press, 1985.

Devlin, Gerald M., and William P. Yarborough. *Paratrooper! The Saga of the U. S. Army and Marine Parachute and Glider Combat Troops During World War II*. New York: St. Martin's Griffin, 1986.

Dilley, Michael F. "Rescue at Cabanatuan." *Behind the Lines*, September–October 1993.

Dilley, Michael F. "Training the Alamo Scouts." *Behind the Lines*, March–April 1995.

Dilley, Michael F. *GALAHAD: A History of the 5307th Composite Unit (Provisional)*. Bennington, VT: Merriam Press, 1996.

Dilley, Michael F. "A Short History of U.S. Airborne Units in World War II." *Tidbits*, August 21, 2012.

Downing, Ben. "A Visit with Patrick Leigh Fermor, Part 2." *The Paris Review Daily*, May 24, 2013.

Durand. "SAARF, POWs, and Tensions between Allies." Posted June 5, 2003. https://www.forum.axishistory.com.

Eisenhower, Dwight D. *Crusade in Europe*. New York: Doubleday & Company, Inc., 1948.

Flanagan, Jr., Edward M. *The Angels: A History of the 11th Airborne Division 1943–1946*. Washington, D.C.: Infantry Journal Press, 1948.

Flanagan, Jr., Edward M. *The Angels: A History of the 11th Airborne Division*. Novato, CA: Presidio Press, 1989.

Flanagan, Jr., Edward M. *11th Airborne*. Paducah, KY: Turner Publishing, 1993.

Flanagan, Jr., Edward M. *Airborne: A Combat History of American Airborne Forces*. New York: Ballantine Publishing, 2002.

Gallagher, Thomas. *Assault in Norway: Sabotaging the Nazi Nuclear Program*. Guilford, CT: Lyons Press, 1975.

Galvin, John R. *Air Assault: The Development of Airmobile Warfare*. New York: Hawthorne Books, 1969.

Gebhardt, James F. *Eyes Behind the Lines: US Army Long-Range Reconnaissance and Surveillance Units*. Fort Leavenworth, KS: Combat Studies Institute Press (Global War on Terrorism Occasional Paper 10), no date.

Gray, David R. "Black and Gold Warriors: U.S. Army Rangers During the Korean War." PhD dissertation, Ohio State University, 1992.

Grimes, William. "Knut Haugland, Sailor on Kon-Tiki, Dies at 92." *New York Times*, January 3, 2010.
Guthrie, Bennett M. *Three Winds of Death: The Saga of the 503rd Parachute Regimental Combat Team in the South Pacific*. Chicago: Adams Press, 1985.
Hasenauer, Heike. "A Special Kind of Hero." *Soldiers*, November 1995.
Haukelid, Hans. *Skis Against the Atom*. London: William Kimber, 1954.
Hicks, Anne. *The Last Fighting General: The Biography of Robert Tyron Frederick*. Atglen, PA: Schiffer Military History, 2006.
Hogan, Jr., David W. "The Evolution of the Concept of the U.S. Army's Rangers, 1942–1983." PhD dissertation, Duke University, 1986.
Hubbell, John G, Andrew Jones, and Kenneth Y. Tomlinson. *P.O.W.: A Definitive History of The American Prisoner-of-War Experience in Vietnam, 1964–1973*. New York: Reader's Digest Press, 1976.
Hughes, Les. "The Alamo Scouts." *Trading Post*, April–June 1986.
Hughes, Les. "The Special Allied Airborne Reconnaissance Force (SAARF)." *Trading Post*, issue unknown, 1991.
Huston, James A. *Out of the Blue: U.S. Army Airborne Operations in World War II*. West Lafayette, IN: Purdue University Studies, 1972.
Hutson, James. "Nathan Hale Revisited." *Library of Congress Information Bulletin*, July–August 2003, 168–172. https://www.loc.gov/loc/lcib/0307-8/hale.html.
Ind, Allison. *Allied Intelligence Bureau: Our Secret Weapon in the War Against Japan*. New York: Curtis Books, 1958.
Jedburgh22. "SAARF." Posted October 31, 2010. https://www.ww2talk.com/index.php.
Jorgenson, Kregg P. "Long Range Recon Patrol: 1901." *Behind the Lines*, November–December 1992.
Kelly, Orr. *From A Dark Sky: The Story of U.S. Air Force Special Operations*. Novato, CA: Presidio Press, 1996.
Kittleson, Galen C., SGM (ret.). Telephone interview, October 1, 1996.
Krueger, Walter. *From Down Under to Nippon: The Story of Sixth Army in World War II*. Nashville: Battery Classics, 1989.
Ladd, James. *Commandos and Rangers of World War II*. London: Macdonald and Jane's, 1978.
Lanning, Michael L. *Inside the LRRPs: Rangers in Vietnam*. New York: Ivy Books, 1988.
Lassen, Don, and Richard K. Schrader. *Pride of America: An Illustrated History of the U.S. Army Airborne Forces*. Missoula, MT: Pictorial Histories Publishing, 1991.
Lipsman, Samuel, ed. *War in the Shadows* (A volume in The Vietnam Experience series). Boston: Boston Publishing, 1988.
Lupyak, Joseph W., SGM (ret.). Telephone interview, September 30, 1996.
MacDougald, N. W. "Who Dares Wins." *Soldier of Fortune*, June 1979.
Martin, David C. "Inside the Rescue Mission." *Newsweek*, July 12, 1982.
Mears, Ray. *The Real Heroes of Telemark: The True Story of the Secret Mission to Stop Hitler's Atomic Bomb*. London: Hodder & Stoughton, 2004.

Mrazek, James E. *The Glider War*. London: Robert Hale, 1975.
Nichols, John, and Tony Rennel. *The Last Escape: The Untold Story of Allied Prisoners of War in Germany, 1944–1945*. United Kingdom: Penguin Books, 2003.
"Obituary—General Paul Aussaresses." *London Telegraph*, December 4, 2013.
Padden, Ian. *U.S. Rangers*. New York: Bantam Books, 1985.
Pugh, Harry. *Rangers: United States Army, 1756 to 1974*. Unpublished manuscript in author's collection.
Raff, Edson D. *We Jumped to Fight*. New York: Eagle Books, 1944.
Rose, Alexander. *Washington's Spies: The Story of America's First Spy Ring*. New York: Bantam Books, 2006.
Rottman, Gordon L. *US Army Rangers & LRRP Units 1942–1987*. London: Osprey Publishing, 1987.
Rottman, Gordon L. *US Army Airborne 1940–90: The First Fifty Years*. London: Osprey Publishing, 1990.
Rottman, Gordon L. *US Airborne Units in the Pacific Theater 1942–45*. London: Osprey Publishing, 2007.
Schemmer, Benjamin F. *The Raid*. New York: Harper and Row, 1976.
Schemmer, Benjamin F. "Requiem for a Warrior." *Soldier of Fortune*, November 1979.
Smith, Bradley F. *The Shadow Warriors: O.S.S. and the Origins of the C.I.A*. New York: Basic Books, 1983.
Smith, Robert Ross. *United States Army in World War II. The War in the Pacific: Triumph in the Philippines*. Washington, D.C.: U.S. Army Center of Military History, 1963.
Stanton, Shelby L. *Vietnam Order of Battle*. Mechanicsburg, PA: Stackpole Books, 1981.
Stanton, Shelby L. *Rangers at War: Combat Recon in Vietnam*. New York: Orion Books, 1992.
Steinberg, Rafael. *Return to the Philippines*. Alexandria, VA: Time-Life Books, 1979.
Ugino, Richard. "Historical Outline: Ranger Units." Unpublished paper in author's collection.
Updegraph, Jr., Charles L. *U.S. Marine Corps Special Units of World War II*. Washington, D.C.: History and Museums Division, HQ US Marine Corps, 1972.
Vandenbrouke, Lucien S. *Perilous Options: Special Operations as an Instrument of U.S. Foreign Policy*. London: Oxford University Press, 1993.
Wells, Jr., Billy E. "The Alamo Scouts: Lessons for LRSUs." *Infantry*, May–June 1989.
Whittaker, Len. *Some Talk of Private Armies*. London: Albanium Publishers, 1984.
Wiggan, Richard. *Operation Freshman: The Rjukan Heavy Water Raid 1942*. London: William Kimber, 1986.
Zedric, Lance Q. "Prelude to Victory: The Alamo Scouts." *Army*, July 1994.
Zedric, Lance Q. *Silent Warriors of World War II: The Alamo Scouts Behind Japanese Lines*. Ventura, CA: Pathfinder Publishing, 1995.
Zedric, Lance Q., and Michael F. Dilley. "Raid on Oransbari." *Behind the Lines*, November–December 1995.
Zedric, Lance Q., and Michael F. Dilley. *Elite Warriors: 300 Years of America's Best Fighting Troops*. Ventura, CA: Pathfinder Publishing, 1996.

Index

1st Airborne Division (Br) 22, 56
1st Airborne Training Battalion 25
1st Battalion, 2nd Regiment, First Special Service Force 52
1st Battalion, 511th Parachute Infantry Regiment 83
1st Canadian Parachute Battalion 44
1st Cavalry Division 64, 81, 88
1st Contingent, SAARF 73
1st Infantry Division 22, 34
1st Marine Parachute Battalion 27
1st Marine Parachute Regiment 27
1st Platoon of the 127th Airborne Engineer Battalion 83
1st Ranger Battalion 26, 92, 97
1st Regiment, First Special Service Force 48, 50
1st Special Forces 96
2nd Airborne Brigade 23
2nd Battalion, 2nd Regiment, First Special Service Force 52
2nd Battalion, 503rd Parachute Infantry Regiment 23
2nd Canadian Parachute Battalion 44
2nd Contingent, SAARF 73
2nd Marine Parachute Battalion 27
2nd Ohio 12, 19
2nd Platoon of the 221st Airborne Medical Company 83
2nd Ranger Battalion 97
2nd Regiment, First Special Service Force 48, 50, 51, 52
3rd Battalion, 511th Parachute Infantry Regiment 83
3rd Contingent, SAARF 73
3rd Infantry Division 95
3rd Marine Parachute Battalion 27
3rd Ranger Battalion 26, 97
3rd Regiment, First Special Service Force 48, 50, 52
4th Marine Parachute Battalion 27
4th Ranger Battalion 26
6th Airborne Division (Br) 22
6th Infantry Division 22, 82, 84
6th Ranger Battalion 65, 83, 105
7th Army 93, 95
11th Airborne Division 24, 25, 29, 81, 82, 83, 88, 93
13th Airborne Division 24, 72
17th Airborne Division 22, 23, 24
21st Amy Group 73
21st Ohio 11, 12, 19
25th Infantry Division 82, 95
29th Infantry Regiment 22
32nd Infantry Division 82, 83
33rd Infantry Division 83
33rd Ohio 12, 18, 19
36th Infantry Division 43, 50, 52
37th Infantry Division 82, 83, 84, 87
54th Troop Carrier Wing 82
75th Infantry Regiment 96, 97
75th Ranger Regiment 97, 98
75th Regimental Combat Team 97
82nd Airborne Division 22, 23, 24, 72, 95, 103

83rd Infantry Division 76
87th Mountain Infantry Regiment 43
88th Glider Infantry Regiment 24
99th Infantry Battalion 26
101st Airborne Division 22, 23, 24, 72, 77, 93
123rd Infantry Regiment 83
126th Infantry Regiment 82
127th Infantry Regiment 83
173rd Airborne Brigade 89
187th Glider Infantry Regiment 24, 82
188th Glider Infantry Regiment 24, 82
193rd Glider Infantry Regiment 24
194th Glider Infantry Regiment 24
317th Troop Carrier Group 84
325th Glider Infantry Regiment 23
326th Glider Infantry Regiment 24
327th Glider Infantry Regiment 23
401st Glider Infantry Regiment 23
433rd Troop Carrier Group 84
474th Infantry Regiment 26
475th Infantry Regiment 97
501st Parachute Infantry Battalion 22, 30
501st Parachute Infantry Regiment 23, 28, 72
502nd Parachute Infantry Battalion 22
502nd Parachute Infantry Regiment 23
503rd Parachute Infantry Battalion 22
503rd Parachute Infantry Regiment 24, 81, 89
504th Parachute Infantry Battalion 22
504th Parachute Infantry Regiment 23, 74, 89
505th Parachute Infantry Regiment 23, 89
506th Parachute Infantry Regiment 23, 28, 29
507th Parachute Infantry Regiment 23, 24
508th Parachute Infantry Regiment 23
509th Parachute Infantry Battalion 23, 24

511th Airborne Signal Company 83
511th Parachute Infantry Regiment 24, 83
511th Parachute Maintenance Company 83
513th Parachute Infantry Regiment 24
515th Parachute Infantry Regiment 24
517th Parachute Infantry Regiment 24
541st Parachute Infantry Regiment 25, 88
542nd Parachute Infantry Battalion 25
542nd Parachute Infantry Regiment 25
550th Airborne Infantry Battalion, a glider unit 24
551st Parachute Infantry Battalion 24
555th Parachute Infantry Battalion 25, 26
711th Airborne Maintenance Company 83
1127th Field Activities Group 102, 103, 104
5307th Composite Unit (Provisional) 97

Abrams, Creighton, General 96, 113
Adairsville, Georgia 16
Adak, Alaska 48
Adams, Paul D., Colonel 47
Addison Point, Florida 28
Admiralty Islands 64
Africa 110
After Action Report 122
Airborne Center, Camp Mackall 25
Airborne Center Training Detachment 25
Airborne Department, Infantry School 30
Airborne School 22, 25
Air Commandos 21
Air Corps 37
Aircraft
 A-1 112, 115, 116
 C-46 Commandos 84, 87

INDEX • 177

C-47 Dakotas 84, 87
C-130—Combat Talon 112, 113, 115
F-4 116
F-105 114, 115, 121
Halifax—bomber 56
HH-3 helicopter, also known as Jolly Green Giant 110, 112, 117, 121
HH-53 helicopter, also known as Super Jolly Green Giant 110, 113, 122
SR-71 113
UH-1 helicopter—also known as Huey and Slick 110
Waco CG-4A gliders 84
Waco CG-13A gliders 84
Aircraft carrier
 USS Hancock 115
 USS Oriskany 115
 USS Ranger 115
Air Force Special Operations Force 106, 109
Alabama 16
Alamo Force—code-name for Sixth Army 61
Alamo Scouts 61, 62, 63, 64, 65, 91, 92, 93, 96, 98, 108, 109
Alamo Scouts teams:
 Chanley Team 65
 Hobbs Team 65, 67
 Littlefield Team 65
 McGowen Team 64
 Rounsaville Team 65
 Sumbar Team 65
 Sumner Team 65
 Thompson Team 65
Alamo Scouts Training Center (ASTC) 61, 91
Alaska 26, 29, 43, 47, 49
Alaska Scouts 48
Albany, New York 1
Alcatraz—POW camp 102

Aleutian Islands 26, 43, 47, 48, 49
Aleutians 61
Alexander, Harold, Field Marshal 78
Allied Intelligence Bureau 61, 91
Allison, John, Lieutenant Colonel 117, 118, 121
Altengrabow, Germany 73, 74, 76
Altoona, Georgia 13, 15
America 3, 9, 101
American Indian 94
Amphibious Scouts 65
Andrews, James J. 11, 12, 13, 15, 16, 17, 18, 19
Anzio, Italy 53
Apache 28
Aparri 81, 82, 83, 84, 88, 89
Ap Lo—POW camp site 102, 104, 105
Army Air Corps 21
Army National Guard 44
Army Quartermaster General 30
Arne Kjelstrup 56
Arnold, Henry H. (Hap), General, Chief of Air Corps 37
ARSOF Memorial Plaza 66
ASTC (Alamo Scouts Training Center) 61, 62, 63, 66, 91, 92, 94, 96
Atlanta, Georgia 11, 12, 18
Augsburg, Germany 93
Aussaresses, Paul, Captain 74
Aux 3 109
Auxiliary Field 3, Eglin Air Force Base—also known as Aux 3 and Duke Field 109
Axis Sally 73

Bailey, Banks, & Biddle 30
Baldwin, Samuel 32
Baldwin, Thomas 32, 33
Balkans 49
Barbara—mock-up of POW compound 110, 112

Bardshar, Frederic A., Vice Admiral 113, 115
Bashur 89
Battery C, 457th Parachute Field Artillery Battalion 83
B Company, 6th Ranger Battalion 105
Beauregard, P. G. T., General 15
Bedford, New York 8
Belgium 37, 40
Bensinger, William, Private 19
Berlin, Germany 73
Berry, Albert 33
Big Shanty, Georgia 12, 13
Blackburn, Donald D., Brigadier General 103, 104, 105, 106, 109, 112, 113, 114, 122
Black Devils—nickname of members of First Special Service Force 53
Black River 116
Blanchard, Jean-Pierre 31, 37
Blueboy 111, 117
Bordeaux, France 55
Bougainville 27
Bradshaw, Frederick W., Lieutenant Colonel 61
Brereton, Lewis H., Major, later Lieutenant General 21, 22, 34, 35
Briar Patch—POW camp 102
BRIEFCASE—SAARF team 74, 75
British Army 1, 3, 7, 8, 9
Britton, Warner A., Lieutenant Colonel 109, 117, 118, 120, 121
Broadwick, Charles 33, 35
Brooks Field, Texas 35
Browne, Elizabeth—wife of Robert Rogers 2
Brown, J., Captain 74
Brussels 73
Buchenwald concentration camp 76
Burgess, Henry, Lieutenant Colonel 83, 87

Cabanatuan, Philippines 65, 108
Cagayan River 82, 84
Cagayan Valley 82, 84
Calgary Highlanders 45
Calhoun, Georgia 16
California 25, 27, 44
Camalaniugan Airfield 83
Cambodia 99
Camo Downes 65
Campbell, William 12, 18, 19
Camp Clairborne, Louisiana 23
Camp Elliot, California 27
Camp Kearney, California 27
Camp Mackall, North Carolina 22, 24, 25
Camp McDonald, 12
Camp Patrick Henry, Virginia 49
Camp Pendleton, California 27
Camp Toccoa, Georgia 22, 28
Cam Rahn Bay 96
Canada 9
Canadian Army 44
Cape Oransbari 65
Casablanca, Morocco 49
CASHBOX—SAARF team 74, 75
Cassino, Italy 43
Cass Station, Georgia 15
Cataldo, Joseph, Lieutenant Colonel 108, 109, 120
Catalina (PBY) 64
Caucasus Mountains 47
Cayley, Sir George 38
Central Intelligence Agency (CIA) 110
Central Office for South Vietnam (COSVN) 99
Chairman of the Joint Chiefs of Staff 104
Chanute, Octave 38, 39
Chattanooga, Tennessee 11, 12, 13, 16, 18
Cherry Point, North Carolina 28
Chief of Staff of the Air Force 106
China 37, 82, 95

INDEX • 179

Chiricahua Apache 28
Choiseul 27
Churchill, Winston 43, 78
CIA (Central Intelligence Agency) 113
CIC (Counter Intelligence Corps) 77
Civil War 100
Clark, Mark, Lieutenant General 43, 49
Claus Helberg 56
Clinton, Henry, General 3
Coast Artillery 44
Cocking, Robert 32
Code of Conduct 101
Colditz Castle, POW camp Oflag IVC 78
Cole, Harold—British deserter 77
Combat Arms Regimental System 96
Combat Talons 115, 116
Command and General Staff College 22, 36
Commandos 92, 93
Committee of Safety 3
Connolly, Robert V., Major 83
Connolly Task Force 84
Contact Teams 71, 72, 73, 74, 75, 76, 77, 78
Continental Congress 3
Coombe-Tenant, Henry, Major 77
Cordillera Central mountains 82, 84
COSVN (Central Office for South Vietnam) 99
Counter-Intelligence Corps (CIC) 77
Cousin, first name not known, S/LT 74, 75
Currahee Mountain 29
Czechoslovakia 93

Dalton, Georgia 18
Da Nang 114
Darby's Rangers 26
Darby, William, Colonel 92
da Vinci, Leonardo 31, 37
Davis, T. J., General 75

D-Day 23, 50
Declaration of Independence 4
Defense Intelligence Agency (DIA) 109, 113, 114
Delaware 7
Demolition Platoon of Headquarters Company, 511th Parachute Infantry Regiment 83
Denton, Jeremiah A., Jr., Captain 101
Deputy Chief of Staff Plans and Operations 103
Detroit, Michigan 1, 2
Deuterium 55
Deuxieme Bureau—French Military Intelligence 55
Doolittle Raid 25, 109
Dorsey, Daniel A., Corporal 19
Dubois, A. E. 30
Duke Field 109

Eagles, John, Captain—company commander in Rogers' Rangers 7, 8
East Germany 93
Eberhardt, Aubrey, Private 29
Edenton, North Carolina 28
Eglin Air Force Base, Florida 106, 109
Eighth Army 88
Eisenhower, Dwight, General 49, 53
England 3, 23, 24, 32, 47, 55, 56, 59, 73
ERASER—SAARF team 74, 75
Europe 24, 77, 82, 112

Fairbairn, Bruce—former Shanghai Police Captain 47
Falmouth, England 55
Fermor, Patrick Leigh, rank not known 77, 78
Fifth Air Force 87
Fifth Army 43, 49

Finger Lake 116
First Airborne Task Force 24, 25, 26
First Allied Airborne Army 21, 73
First Special Service Force 24, 26, 29, 43, 44, 45, 46, 47, 48, 49, 50, 51, 52, 53
Fleet Marine Force 28
Florida 48, 93, 106
Flushing, New York 5
Fonda, Jane 100
Forshall, Sam, Major 74
Fort Belvoir, Virginia 102
Fort Benning, Georgia 22, 25, 29, 88, 93, 97, 105
Fort Bragg, North Carolina 66, 105, 106, 108
Fort Campbell, Kentucky 93, 96
Fort Eban Emael 40
Fort Ethan Allen, Vermont 48
Fort Gordon, Georgia 97
Fort Knyphausen 8
Fort Leavenworth, Kansas 22
Fort Lewis, Washington 43, 97
Fort Michilimackinac 3
Fort Pierce, Florida 48
Fort William Henry Harrison, Helena, Montana 26, 44, 45, 47
Fourteenth Army 81, 82
Fourth Allied POW Wing 123
France 23, 24, 25, 26, 34, 49, 72
Frederick, Robert T., Lieutenant Colonel, later Major General 24, 44, 45, 47, 48, 50, 51, 52
French and Indian War 1, 2, 5
French Resistance 72
French Revolution 32
Fuller, William—conductor 13, 15, 16, 17, 18

Garnerin, Andre-Jacques 31, 32
Garnerin, Jeanne-Genevieve, wife of Andre-Jacques 32

Gavutu 27
General—locomotive 12, 13, 16, 17, 18
Geneva Accords on Treatment of Prisoners of War 100
Geneva Convention 75
Georgia 11, 12, 22, 27, 93, 97, 105
German Winter Line 43, 49, 53
Germany 22, 24, 36, 37, 39, 40, 55, 58, 71, 73, 77, 82, 93, 95
Geronimo 28, 29
Gestapo 59, 77
get on the walls—communicating using the tap code 100
ghost airborne divisions 26
Glider Detachment (USMC) 28
Glider Group 71 (USMC) 28
Glider Infantry Battalions (GIB) 23
Glider Infantry Regiments (GIR) 23, 82
Glover, John, General 8
Golden Gate Park 33
Green, first name not known, Major 8
Greenleaf 111, 117
Greenville, Connecticut 5
Grenada 97
Guadalcanal 27
Gypsy Task Force 81, 83, 84, 87, 88, 89

Hai Phong 104, 116
Hai Phong Harbor 113
Hale, Nathan, Captain 5
Halle, Germany 77
Hampton Roads Port of Embarkation 49
Hanoi 99, 102, 115
Hanoi Hilton—POW camp Hoa Lo 102
Hardanger plateau 56
Haslemere, Waverly 78
Haslet, John, Colonel 7, 8
Haugland, Knut 56, 58, 59
Haukelid, Knut 57
Hawaii 113
Hawkins, Martin J., Corporal 12, 19

Headquarters and Service Squadron 71 28
Headquarters, Marine Corps 27
Heartbreak Hotel 100
heavy water 55, 56, 57, 58, 59
Hegdahl, Douglas B., Seaman 101, 102
Helena, Montana 26, 44
Hempstead, Stephen, Sergeant 5
Heraldry Section, Quartermaster Corps 45
Herman Goering Division 49
Heyerdahl, Thor 59
Hiroshima 88
Hitler, Adolf 55, 56, 71
Hoa Lo—POW camp 100, 102, 123
Hobbs Team 65, 67
Hobbs, Woodrow H., Kieutenant—leader of Hobbs Team 65, 67
Hoffman, E. L., Major 35
Hon Tre Island 95
Hopkins, Joseph, Captain 2
Hopkins' Rangers 2
Howe, William, General 4, 5, 8, 9
Huey 110

I Airborne Corps (Br) 22
Idland, Kasper 57
Igorot headhunters 103
Independent Grenadier Company—Polish 72
Indian Scouts 29, 45
Infantry School, Fort Benning 30, 93, 105
Innes, Alexander—Inspector General of Provincial Forces 9
Inspector General of Provincial Forces 9
Intelligence Target Folders 102
Iraq 89
Irish Guards 77
Irvin, Leslie L. 35
Italy 23, 26, 43, 47, 49, 95

Japan 65, 82, 88
Japen Island 65
JCTG—Joint Contingency Task Group 106
Jedburghs 21, 72
Jefferson Barracks, Missouri 33
Jens-Anton Poulsson, 56
Johnson, Lyndon B. 99
Joint Contingency Task Group (JCTG) 106
Jolly Green Giant 110

Kahili—a Japanese airfield 27
Kalen, Herbert, Major 117, 121
Kara—a Japanese airfield 27
Kayser, Fredrik 57
Kelly Field, Texas 35
Kemmer, Thomas J., Master Sergeant 120
Kentucky 93
King George III 3
King's Rangers 9
Kingston, Georgia 15
Kiska, Alaska 26, 43, 47, 48
Kiska Harbor 48
Kiska Island 61
Kissinger, Henry A. 112
Kittleson, Galen C., Master Sergeant 108, 109, 117
Knight, William, Private 11, 12, 13, 15, 19
Kon Tiki expedition 59
Korea 93
Korean War 93, 96, 97
Krueger, Walter, Lieutenant General, Commander, Sixth Army 61, 65, 67, 68, 82, 83, 84, 91, 92, 108
Kuhberge, Germany 76

Lackey, John, Colonel 84
Lacy Hotel 12

la Difensa—Italian mountain 43, 49, 51, 52
Laird, Melvin 114, 115
Lake Constance 77
Lake Tinnsjo 58
Langenstein, Germany 76
Langley, Samuel 38
Laos 103, 105, 115, 116, 122
Laotian Army 103
la Remetanea—Italian mountain 43, 49, 52
Laurent, Francois 38
LAW 121
Lee, Charles, General 8
Lee, William C., Major 37
Lenormand, Louis-Sebastien 31
Leyte 65
Lilienthal, Gustav 38
Lilienthal, Otto 38, 39
Lillyman, Frank, Captain 77
Lipa airfield 81, 82, 87, 88
Llewellyn, Samuel, Corporal 12, 19
London 2, 3, 4, 9, 47, 57
Long Island 5
Long Range Patrols (LRPs) 65, 95
Long Range Reconnaissance Patrol (LRRP) 65, 93, 95, 96
Long Range Surveillance Units 66
Lord Harewood 78
Los Baños POW camp 81
Los Negros 64
Louisiana 23
Lounsbury, William—recruiter for Rogers' Rangers 5
LRRP (Long Range Reconnaissance Patrol) 63, 93, 95, 96
Lungo—Italian mountain 43
Lupyak, Joseph W., Master Sergeant 118
Luzon 64, 65, 81, 82, 83, 88, 103, 108
Lynch, George A., General, Chief of Infantry 36

Maan River 57
MacArthur, Douglas, General 61, 88, 91
MACV (Military Assistance Command Vietnam) 95, 96
MACV Recondo School 66, 96
Malay Peninsula 81
Mamaroneck, New York 7, 8
Manhattan 5
Manila 65, 82
Manor, Leroy J., Brigadier General 106, 108, 109, 112, 113, 114, 122, 123
Marblehead Mariners 8
Marietta, Georgia 12, 18
Marine Corps Air Station, Eagle Mountain Lake, Texas 28
Marine Corps Air Station, Parris Island, South Carolina 28
Marine Glider Squadron 711 28
Marshaling Area Control Officer 121
Marshall, George C., General, Army Chief of Staff 36, 37
Martin bomber 35
Maryland 3, 7
Mason, Elihu H., Sergeant 19
Mayer, Edward E., Colonel 103, 104, 105, 106, 113
McCook Field, Ohio 35
McEntee, Ducat, Colonel 88
McGowen, John R. C., Lieutenant— leader of McGowen Team 64, 67
McQueen, John G., Lieutenant Colonel 45, 47
Meadow, Richard J., Captain 105, 108, 109, 110, 117, 118, 120, 121
Meadows' team—also known as Blueboy 111, 118, 121
Memphis, Tennessee 11
Mercedes 77
Merrill's Marauders 97
Metz, Germany 21, 22, 34

Mignano Gap, Italy 43
Military Assistance Command Vietnam (MACV) 113
Military decoration
   Basic Parachutist Insignia (Jump wings) 30, 31
   Glider Badge 30
   Medal of Honor 18, 19, 94
   Navy and Marine Corps Parachutist Insignia 31
   War Cross with Sword 59
Miller, Walter L., Staff Sergeant 118, 121
Millett, Lewis, Major 94
Missouri 33
Mitchell, William P. (Billy), Colonel, later Brigadier General, 21, 34, 35
Mitchel, Ormsby, Major General 11, 16, 18
Monkey Mountain 114, 122
Montgolfier, Jacques 37
Montgolfier, Joseph 37
Moorer, Thomas, Admiral 105, 106, 113, 114
Morse code 100, 101
Mount Malepunyo 81
Mount Rainier 44
MPB (Marine Parachute Battalion) 27

Nadzab, New Guinea 81
Nagasaki 88
Nakhon Phanon 115
Naples, Italy 49
Nashville, Tennessee 11
National Military Command Center 114
National Security Advisor 112
National Security Agency 104
Naval Air Station Lakehurst 27
Navy Scouts and Raiders School 48
Nellist Team of Alamo Scouts 65, 108

New England 3
New Georgia Group 27
New Guinea 24, 47, 65, 81, 108
New Hampshire 4
New Hampshire House of Representatives 4
New Jersey 27
New Orleans 32
New River, North Carolina 27
New Rochelle, New York 7
New York City, New York 4
Nha Trang, Vietnam 96
Nichols, J. S., Brigadier 72, 75
Nickerson, David S., Staff Sergeant 118, 121
Nixon Administration 101
Nixon, Richard M. 99, 112, 113
No. 1 Parachute Training School 73
Noemfoor Island 89
Norfolk, Virginia 47
Norman, Oklahoma 28
Norsk Hydro 55, 56, 57, 58
North Africa 23
North America 25
North Carolina 22, 27, 28, 105
North Vietnam 99, 100, 103, 104, 108, 110, 114, 116, 122
Northwest Passage 3
Norwalk, Connecticut 5
Norway 26, 37, 43, 47, 55, 56, 57, 59
NVA (North Vietnamese Army) 95

Ochernal, first name not known, Colonel 75, 76
Office of Strategic Services (OSS) 21, 61, 91
Oflag IVC, also known as Colditz Castle 78
Ohio National Guard 82
Okinawa 88, 97

O'Neill, D. M.—Shanghai police officer, developer of the O'Neill System 47
O'Neill System—hand-to-hand fighting technique 26, 47
Operation *Freshman* 56, 57
Operation *Grouse* 56, 57
Operation *Gunnerside* 57
Operation *Ivory Coast* 106, 110
Operation *Kingpin* 113
Operation *Polar Circle* 104
Operation *Popeye* 122
Operation *Raincoat* 43
Operation *Swallow*—rename for Operation *Grouse* 57, 58
Operation *Varsity* 73
Operation *Vicarage* 72
Operation *Violet* 74, 75, 76, 77
Operations Groups 21
Oran, Algeria 49
Oransbari, New Guinea 108
Ordnance Department 47
Ormoc 65
Oslo, Norway 55, 58
OSS—Office of Strategic Services 61, 72, 73, 74, 75, 91

Pacific 24, 27, 28, 88, 113
Pacific Fleet Headquarters 48
Pacific Northwest 25
Pacific Ocean 59
Paige Field, Parris Island, South Carolina 28
Panzer Grenadier Division 49
Parachute Infantry Battalions (PIB) 22
Parachute Infantry Regiments (PIR) 23, 89
Parachute Test Platoon (PTP) 22, 29
Paris 31, 55, 77, 101
Parrott, Jacob, Private 18, 19
Patton, George S., General 78, 123
Pearl Harbor, Hawaii 48

Pegun Island 65
PENCIL—SAARF team 74
PENNIB—SAARF team 74
Pentagon 114
Pershing, John J., General 34, 35
Perth, Scotland 55
Petrie, George W., Lieutenant 117, 120
Phantom Reconnaissance Force 78
Philadelphia 3, 30
Philippine 11th Infantry, a guerrilla unit 83
Philippines 24, 25, 81, 82, 103, 114
PIB (Parachute Infantry Battalions) 22, 23, 24, 25, 30
Pilatre de Rozier, Jean-Francois 38
PIR (Parachute Infantry Regiments) 23, 24, 25, 28, 74
Pittenger, William, Corporal 19
Plan Green 117, 118
Ploesti, Rumania 47
Plutonium-239 55
Pontiac's Conspiracy 1
Porter, John R., Private 12, 19
POWs—Prisoners of War 71, 74, 75, 76, 78, 81, 94, 101, 102, 103, 108, 110, 113, 114, 116, 118, 122, 123
Project Plough 43, 44, 45, 47
PT boat 64, 65
PTP (Parachute Test Platoon) 22
Putnam, Rufus 7
Pyke, Geoffrey—British scientist 43

Quartermaster Corps 45, 47
Queen's Royal American Rangers 2, 4, 9

Radio Berlin 73
RAF Great Dunmow airfield 75
Ranger/Airborne companies 93, 97
Ranger Department, Infantry School 105

Rangers 1, 4, 5, 7, 8, 9, 92, 105
Ranger School 66, 93, 94, 96
Raymond, J. E., Colonel 72
Recondo 91, 94, 95, 96, 97
Recondo Schools 97
Reconnaissance and Commando 94
Reddick, William H. H., Corporal 19
Redwine 111, 117
Reed, Miles, rank not known 78
Resaca, Georgia 17
Rescue Combat Air Patrol 115
Rhine River 73
Ringgold, Georgia 18
Ringway, England 73
Rjukan 55
Robertson, Samuel, Private 19
Robinson, Bill L., Lieutenant 110, 111
Rockpile—POW camp 102
Rogers, James, brother of Robert Rogers 9
Rogers' Rangers 1, 4, 7, 8
Rogers, Robert, Major 1, 2, 3, 4, 5, 7, 8, 9
Rome, Italy 53
Ronneberg, Joachim 57
Roosevelt, Franklin D., President of the United States 37
Ross, Marion A., Sergeant 19
ROTC (Reserve Officer Training Corps) 97
Rotondo—Italian mountain 43
Route 3 83
Route 5 82, 83, 87
Rowe, James N. (Nick) 109
Rumania 47
Russia 22, 36, 40
Russian army 76

SAARF—Special Allied Airborne Reconnaissance Force 72, 73, 74, 75, 76, 77, 78, 79
SACSA—Special Assistant to the Chairman (of the Joint Chiefs of Staff) for Counterinsurgency and Special Activities 103
Saigon 113
Salerno, Italy 89
San Diego, California 27
San Francisco, California 33, 48
Sarmi Harbor 65
SAS—Special Air Service 72
Saxony, Germany 78
Scotland 55, 92
Scott, John M., Sergeant 19
SEALINGWAX—SAARF team 74, 76
Secretary of Defense 114
Service Battalion, First Special Service Force 45, 52
Service Company, 511th Parachute Infantry Regiment 83
SF Hydro—ferry 58, 59
Shadrack, Charles P., Private 11, 19
SHAEF—Supreme Headquarters Allied Expeditionary Force 71, 72, 73, 78
Shanghai 47
Shanghai Police Department 47
Shimbu Group, Fourteenth Army 81
Shobu Group, Fourteenth Army 82, 84, 88
Sicily, Italy 23
Sierra Madre mountains 82
Sierra Nevada—mountain range 44
Simcoe, James G., Colonel 9
Simons, Arthur D. (Bull), Colonel 105, 106, 108, 109, 110, 112, 113, 115, 117, 118, 120, 121, 122, 123
Simons' team—also known as Greenleaf 111, 117
Sixth Army 61, 63, 65, 82, 88, 91, 105, 108
Skidrow—POW camp 102
Slavens, Samuel, Private 19

slick 110
Smith, James, Private 12, 19
Smokejumpers 25
SOE—Special Operations Executive 56, 57, 58, 72, 73, 74, 75, 77
SOG—Studies and Observation Group or Special Operations Group 103, 105, 106
Solomon Islands 27
Song Con River 106, 116, 117
Son Tay—POW camp site 99, 102, 103, 104, 105, 106, 109, 110, 112, 113, 114, 115, 116, 117, 118, 120, 122, 123
South Amboy, New Jersey 4
South America 105
South Carolina 1, 28
South China Sea 95
Southeast Asia 110
Southern European Task Force 95
Southern France 25, 53
South Vietnam 109, 113, 114, 122
South Wales Borderers 74
Southwark, England 9
Southwest Pacific Theater 81, 91
Special Air Service (SAS) 78
Special Allied Airborne Reconnaissance Force (SAARF) 21, 71, 72
Special Assistant to the Chairman (of the Joint Chiefs of Staff) for Counterinsurgency and Special Activities (SACSA) 103
Special Forces 64, 66, 93, 96, 97, 103, 105, 106, 108, 109, 112, 118, 121
Special Operations Command 66
Special Operations Division—of SACSA 103
Special Operations Executive (SOE) 56
Special Operations Group—also known as Studies and Observation Group (SOG) 103

Special Reconnaissance Unit of Sixth Army 91
Special Troops, 511th Parachute Infantry Regiment 83
Spencer, Herman, Master Sergeant 118
Stalag XIA—POW camp 74, 75
Stamford, Connecticut 5
Stanton, Edwin 18
Stevens, Alabama 16
Stevens, Leo 33
Stirling, David, Lieutenant Colonel 78
Storhaug, Hans 57
Stratton, Richard A., Lieutenant Commander 101
Stromsheim, Birger 57
Studies and Observation Group (cover name for Special Operations Group) (SOG) 103
Sunningdale golf course 73
Super Jolly Green Giant 110
Supreme Headquarters Allied Expeditionary Force (SHAEF) 71
Surrey 73
Sweden 56, 58
Swing, Joseph, Major General 81
Syndor, Elliot P., Lieutenant Colonel 105, 106, 108, 109, 110, 115, 117, 118, 120
Syndor's team—also known as Redwine 111, 117, 118, 121

Ta Khli Royal Thai Air Base 113, 115
Tanambogo 27
tap code—communication method between POWs 100, 102
Task Force Connolly 83
Tay Ninh Province 89
Telemark 55
Tennessee 11, 12
Tet Offensive 99
Texas—locomotive 16, 17

Thailand 104, 110, 113, 114, 115, 122
The Commando Order 56
The Netherlands 23, 24, 37
the Zoo—POW camp 102
Third Army 78
Tilton, Georgia 17, 18
Ton Kin Gulf 113
Treaty of Versailles 36, 39
Troop Carrier 21
Trumbull, Jonathan—Governor of Connecticut 5
Tunis 78
Tunnel Hill, Georgia 18
Typhoon Patsy 113, 114

U Dorn Royal Thai Air Base 104, 115
United States 55
Uranium 55
U.S. Air Force 102
U.S. Army 22, 25, 26, 28, 31, 33, 35, 36, 43, 44, 45, 66, 77, 95, 105
US Army Air Forces 58
U.S. Forest Service 25
U.S. Marine Corps 27, 47
U.S. Military Academy 95
U.S. Navy 26, 62, 101, 116
U.S. Patent and Trademark Office 30

Vandeomoear Island 65
V Corps 95
VC (Viet Cong) 95
Vella Lavella, New Georgia Group 27
Vermont 48
Vietnam 63, 65, 89, 95, 96, 97, 99, 101, 103, 105, 106, 108
Vietnam War 97, 99, 106
Vigan 83
VII Corps 95
Virginia 7, 8, 47, 49, 102

Walther, Udo, Captain 120

War Department 30, 44, 45
Warfield, first name not known, Captain 74
Wargeo Island 65
Washington (DC) 18, 46, 91, 106, 109, 112, 113, 114
Washington, George, General 3, 4, 5, 8
Washington (state) 43, 97
Weapons
    105mm howitzers 83
    anti-aircraft batteries 99, 114
    Bayonets and knives 51
    booby traps 92
    flame throwers 87
    Johnson light machine guns 47
    Light Anti-tank Weapon (LAW) rockets 121
    M-60 machine guns 110, 117
    machine guns 51
    mortar shells 52
    Rockeyes—cluster ammunition 115
    RS—an explosive 47
    SAM (surface-to-air missiles) 114, 116, 118, 121
    submachine gun 58
    surface-to-air missiles (SAM) 99
Weasel—propellor-driven sled 43, 44
Welsh Guards 77
Wentworth, Surrey 73
Westmoreland, William, General 93, 94, 95, 96
West Point, New York 95
Wheeler, Earle, General 104
White, Horton V., Colonel, G-2, Sixth Army 37, 61, 103, 105, 113
White House 37, 113
White Star—Special Forces teams operating in Laos 103, 105
William B. Smith—locomotive 15, 16
Wilson, George D., Private 19
Wollam, John, Private 19
Wood, Mark, Private 19
World War I 21, 33, 39

World War II 21, 31, 55, 81, 88, 91, 93, 96, 97, 105, 106, 108
Worrall, Phillip, Major 74, 75, 76
Wright, Larry M., Technical Sergeant 117
Wright, Orville 38, 39
Wright, Wilbur 38

XVIII Airborne Corps 22, 24, 95, 105
XVIII Corps 24

Yamashita, Tomoyuk General (the Tiger of Malaya) 81, 88
Yankee Station 113, 115, 116
Yarborough, William P., Captain, later Lieutenant General 30
Yonah—small locomotive 13, 15

Zerbst, Germany 76